ENERGY SECTOR STANDARD
OF THE PEOPLE'S REPUBLIC OF CHINA

中华人民共和国能源行业标准

Code for Design of Hoists for Hydropower Projects
Part 1: Code for Design of Fixed Wire Rope Hoists

水电工程启闭机设计规范
第1部分：固定卷扬式启闭机设计规范

NB/T 10341.1-2019

Replace DL/T 5167-2002

Chief Development Department: China Renewable Energy Engineering Institute

Approval Department: National Energy Administration of the People' s Republic of China

Implementation Date: July 1, 2020

China Water & Power Press
中国水利水电出版社
Beijing 2024

图书在版编目（CIP）数据
水电工程启闭机设计规范. 第1部分，固定卷扬式启闭机设计规范 : NB/T 10341. 1-2019 = Code for Design of Hoists for Hydropower Projects Part 1: Code for Design of Fixed Wire Rope Hoists (NB/T 10341. 1-2019) : 英文 / 国家能源局发布. -- 北京 : 中国水利水电出版社, 2024. 10. -- ISBN 978-7-5226-2798-4
Ⅰ. TV664-65
中国国家版本馆CIP数据核字第2024E4R143号

ENERGY SECTOR STANDARD
OF THE PEOPLE'S REPUBLIC OF CHINA
中华人民共和国能源行业标准

Code for Design of Hoists for Hydropower Projects
Part 1: Code for Design of Fixed Wire Rope Hoists
水电工程启闭机设计规范　第1部分：固定卷扬式启闭机设计规范

NB/T 10341.1-2019

Replace DL/T 5167-2002

（英文版）

Issued by National Energy Administration of the People's Republic of China
国家能源局　发布
Translation organized by China Renewable Energy Engineering Institute
水电水利规划设计总院　组织翻译
Published by China Water & Power Press
中国水利水电出版社　出版发行
Tel: (+ 86 10) 68545888　68545874
sales@mwr.gov.cn
Account name: China Water & Power Press
Address: No.1, Yuyuantan Nanlu, Haidian District, Beijing 100038, China
http: //www.waterpub.com.cn
中国水利水电出版社微机排版中心　排版
北京中献拓方科技发展有限公司　印刷
184mm×260mm　16 开本　7 印张　221 千字
2024 年 10 月第 1 版　2024 年 10 月第 1 次印刷
Price（定价）: ¥ 350.00

Introduction

This English version is one of China's energy sector standard series in English. Its translation was organized by China Renewable Energy Engineering Institute authorized by National Energy Administration of the People's Republic of China in compliance with relevant procedures and stipulations. This English version was issued by National Energy Administration of the People's Republic of China in Announcement [2023] No. 5 dated October 11, 2023.

This version was translated from the Chinese Standard NB/T 10341.1-2019, *Code for Design of Hoists for Hydropower Projects—Part 1: Code for Design of Fixed Wire Rope Hoists*, published by China Water & Power Press. In the event of any discrepancy in the implementation, the Chinese version shall prevail.

Many thanks go to the staff from the relevant standard development organizations and those who have provided generous assistance in the translation and review process.

For further improvement of the English version, any comments and suggestions are welcome and should be addressed to:

China Renewable Energy Engineering Institute
No. 2 Beixiaojie, Liupukang, Xicheng District, Beijing 100120, China
Website: www.creei.cn

Translating organization:

POWERCHINA Zhongnan Engineering Corporation Limited

Translating staff:

LIU Xiaofen LI Qian WANG Hongfang HU Caishi
GUO Yulan YE Yuxin

Review panel members:

GUO Jie	POWERCHINA Beijing Engineering Corporation Limited
QIAO Peng	POWERCHINA Northwest Engineering Corporation Limited
LI Zhongjie	POWERCHINA Northwest Engineering Corporation Limited

JIN Xiaohua	POWERCHINA Huadong Engineering Corporation Limited
LIANG Hongli	Shanghai Investigation, Design & Research Institute Co., Ltd.
ZHANG Qingjun	China Gezhouba Group Machinery & Ship Corporation Limited
JIA Haibo	POWERCHINA Kunming Engineering Corporation Limited
TAO Yundong	Guangdong Hydropower Planning & Design Institute Corporation Limited
LIN Zhaohui	China Renewable Energy Engineering Institute
LI Shisheng	China Renewable Energy Engineering Institute

National Energy Administration of the People's Republic of China

翻译出版说明

本译本为国家能源局委托水电水利规划设计总院按照有关程序和规定，统一组织翻译的能源行业标准英文版系列译本之一。2023 年 10 月 11 日，国家能源局以 2023 年第 5 号公告予以公布。

本译本是根据中国水利水电出版社出版的《水电工程启闭机设计规范　第 1 部分：固定卷扬式启闭机设计规范》NB/T 10341.1—2019 翻译的，著作权归国家能源局所有。在使用过程中，如出现异议，以中文版为准。

本译本在翻译和审核过程中，本标准编制单位及编制组有关成员给予了积极协助。

为不断提高本译本的质量，欢迎使用者提出意见和建议，并反馈给水电水利规划设计总院。

地址：北京市西城区六铺炕北小街 2 号
邮编：100120
网址：www.creei.cn

本译本翻译单位：中国电建集团中南勘测设计研究院有限公司

本译本翻译人员：刘小芬　李　倩　王洪方　胡彩石　郭玉兰　叶雨欣

本译本审核人员：

郭　洁　中国电建集团北京勘测设计研究院有限公司

乔　鹏　中国电建集团西北勘测设计研究院有限公司

李仲杰　中国电建集团西北勘测设计研究院有限公司

金晓华　中国电建集团华东勘测设计研究院有限公司

梁洪丽　上海勘测设计研究院有限公司

张庆军　中国葛洲坝集团机械船舶有限公司

贾海波　中国电建集团昆明勘测设计研究院有限公司

陶云冬　广东省水利电力勘测设计研究院有限公司

林朝晖　水电水利规划设计总院

李仕胜　水电水利规划设计总院

国家能源局

Announcement of National Energy Administration of the People's Republic of China
[2019] No. 8

National Energy Administration of the People's Republic of China has approved and issued 152 energy sector standards including *Code for Operating and Overhauling of Excitation System of Small Hydropower Units* (Attachment 1), and the English version of 39 energy sector standards including *Code for Safe and Civilized Construction of Onshore Wind Power Projects* (Attachment 2).

Attachments: 1. Directory of Sector Standards

2. Directory of English Version of Sector Standards

National Energy Administration of the People's Republic of China

December 30, 2019

Attachment 1:

Directory of Sector Standards

Serial number	Standard No.	Title	Replaced standard No.	Adopted international standard No.	Approval date	Implementation date
…						
16	NB/T 10341.1-2019	Code for Design of Hoists for Hydropower Projects Part 1: Code for Design of Fixed Wire Rope Hoists	DL/T 5167-2002		2019-12-30	2020-07-01
…						

Foreword

According to the requirements of Document FGBGY [2006] No. 1093 issued by the General Office of National Development and Reform Commission of the People's Republic of China, "Notice on Releasing the Development Plan of Sector Standards in 2006", and after extensive investigation and research, summarization of practical experience, consultation of relevant standards in China, and wide solicitation of opinions, the drafting group has prepared this code.

The main technical contents of this code include: basic requirements, design principles and requirements, loads, materials, mechanical design, structural design, electrical design, and safety.

The main technical contents revised are as follows:

This code replaces the content regarding the design of fixed wire rope hoists specified in DL/T 5167-2002, *Design Specifications for Gate Hoist in Hydropower and Water Resources Projects*.

— Adding the content about the torque calculation for motors.

— Adding the content about safety brakes.

— Adding the calculation of time and acceleration/deceleration for starting/braking the hoisting mechanism.

— Adding the calculation of maximum static tension of the wire rope in operation.

— Adding the content about broken line groove drums.

— Adding the allowable stress formulae in the structural calculation.

— Adding the chapter "Safety".

— Modifying the principle of hoist duty classification.

— Modifying the chapter "Electrical Design".

— Revising the requirements for structural member materials.

National Energy Administration of the People's Republic of China is in charge of the administration of this code. China Renewable Energy Engineering Institute has proposed this code and is responsible for its routine management. Energy Sector Standardization Technical Committee on Hydropower Steel Structures and Hoists is responsible for the explanation of specific technical contents. Comments and suggestions in the implementation of this code should be addressed to:

China Renewable Energy Engineering Institute
No. 2 Beixiaojie, Liupukang, Xicheng District, Beijing 100120, China

Chief development organization:

POWERCHINA Zhongnan Engineering Corporation Limited

Participating development organization:

Sinohydro Jiajiang Hydraulic Machinery Company Limited

Chief drafting staff:

CHEN Huichun	YU Xiquan	LI Qijiang	HU Caishi
WANG Hongfang	YAO Gang	YUAN Changsheng	HUANG Wenli
JIANG Lixin	GUO Xihong	WU Sigou	GONG Zhaohui
FANG Zuowei			

Review panel members:

GONG Jianxin	LIN Zhaohui	CHEN Hong	HU Baowen
FANG Hanmei	YAO Changjie	CHEN Xia	LU Wei
LI Lili	ZHAO Fuxin	WU Quanben	LUO Wenqiang
LIAO Yongping	LIU Shuyu	JIN Xiaohua	LONG Zhaohui
YANG Qinghua	WU Xiaoning	PIAO Ling	LI Shisheng

Contents

1 General Provisions

1.0.1 This code is formulated with a view to standardizing the design of fixed wire rope hoists for hydropower projects.

1.0.2 This code is applicable to the design of motor-driven fixed wire rope hoists for operating the gates or trash racks of hydropower projects.

1.0.3 In addition to this code, the design of fixed wire rope hoists for hydropower projects shall comply with other current relevant standards of China.

2 Terms

2.0.1 fixed wire rope hoist

mechanical equipment mounted above the slot top for hoisting a gate or trash rack, where the pulley block or sling is moved up or down by the wire rope wound on the drum rotated by the motor-driven gear transmission

2.0.2 incense coil hoist

one type of fixed wire rope hoist, where each wire rope is wound in multiple layers on a plumb plane

2.0.3 lifting force

hoisting force required to open a gate through slings

2.0.4 holding force

hoisting force applied to the slings to control the gate lowering speed during closing

2.0.5 supporting brake

normally closed mechanical brake installed on the high-speed shaft of a reducer for braking and stopping the hoist

2.0.6 safety brake

normally closed mechanical brake installed on the drum for braking the hoist

2.0.7 spiral groove drum

drum with rope grooves spirally distributed at each pitch

2.0.8 broken line groove drum

drum in which grooves parallel to and intersecting with the plane perpendicular to the drum axis are alternately distributed at each pitch

2.0.9 closed gear transmission

well lubricated transmission in which the gear pairs from the motor to the drum shaft are all in the enclosed box

2.0.10 duty class

hoist characteristic determined based on the state of loading, utilization, and purpose of the gate

3 Basic Requirements

3.0.1 The utilization class, loading state, and duty class of a hoist shall be in accordance with Tables 3.0.1-1, 3.0.1-2 and 3.0.1-3, respectively; the duty class of hoists for different types of gates should be in accordance with Table 3.0.1-4.

Table 3.0.1-1 Utilization class

Utilization class	Total design working hours (h)	Remarks
T1	800	Infrequently used
T2	1600	
T3	3200	Moderately used
T4	6300	Frequently used

Table 3.0.1-2 Loading state

Loading state	Description
L1	Rarely lift the maximum operating load, normally lift light loads
L2	Occasionally lift the maximum operating load, normally lift moderate loads
L3	Frequently lift the maximum operating load, normally lift heavy loads

Table 3.0.1-3 Duty class

Loading state	Utilization class			
	T1	T2	T3	T4
L1	Q1	Q1	Q1	Q2
L2	Q1	Q1	Q2	Q3
L3	Q1	Q2	Q3	Q4

Table 3.0.1-4 Duty classes of hoists for different types of gates

Type of gate	Lift H (m)	Duty class
Bulkhead gate	< 40	Q1
	≥ 40	Q1 - Q2
Emergency gate	< 40	Q2
	≥ 40	Q2 - Q3
Service gate	< 40	Q2 - Q3
	≥ 40	Q3 - Q4

3.0.2 The following data shall be collected for the design of hoists:

1 Gate operating requirements, such as operating conditions and water filling mode.

2 Lifting force, holding force, lifting height, opening speed, closing speed, and synchronism deviation between two lifting points.

3 Layout of hydraulic structures, dimensions of gates and slots, and relevant dimensions and requirements for connection between gates and hoists.

4 Electric drive, control modes and interface requirements.

5 Environmental conditions such as hydrology, meteorology, sediment, water quality, and elevation.

6 Self-weights of gates, tie rods, balance beams, etc.

7 Conditions and requirements for manufacturing, transportation, and installation.

8 Seismic data.

9 Requirements for service power supply, emergency power supply, and control power supply.

10 Contract documents and other special requirements.

3.0.3 Series parameters of hoisting force, lift and hoisting speed should be in accordance with Appendix A of this code.

3.0.4 Hoists shall be provided with safety protection and detection devices for mechanical and electrical equipment, and online monitoring devices should be provided for important hoists.

3.0.5 Hoists shall be protected against high temperature, rain, freezing, moisture, corrosion, wind, sand, and lightning, and shall be provided with ventilation and dehumidification, according to the operating environment.

3.0.6 The fatigue strength need not be calculated for the structural members of the hoist.

3.0.7 The hoist transportation unit should not exceed the maximum overall dimensions and weight specified for transportation and limited by the site installation conditions, and shall have adequate stiffness.

3.0.8 Hoists operating in regions with a seismic intensity of Ⅶ or above shall be verified for seismic safety, and seismic measures shall be taken.

4 Design Principles and Requirements

4.0.1 The design of hoists shall meet the requirements for safety and serviceability, reliable operation, technological advancement, economic rationality, easy maintenance, landscaping, workplace safety, energy conservation, environmental protection, structural coordination, and friendly human-machine interface.

4.0.2 The type selection of hoists shall be determined through a techno-economic demonstration according to the layout of hydraulic structures, gate type, number of orifices, operation and time requirements, and should meet the following requirements:

1 For the service gates of water-release system and sediment-flushing system, each gate should be provided with a hoist, and the fixed wire rope hoist may be selected through type comparison.

2 For the emergency gates and bulkhead gates serving fewer orifices, the fixed wire rope hoist may be selected through type comparison.

3 For the quick-acting shutoff gates at the power intake, the fixed wire rope hoist may be selected through type comparison.

4 For the closure gates of construction diversion works, the fixed wire rope hoist should be selected.

4.0.3 A room should be provided for fixed wire rope hoists. The space for maintenance shall be reserved in the hoist room, and passageways, which should be greater than 0.8 m in width, shall be reserved between the hoist and the walls. Maintenance facilities shall be provided in rooms for large hoists. Outdoor hoists shall be provided with a removable cover. For the fixed wire rope hoists used in severe cold regions and operating in winter, thermal insulation facilities shall be provided in the hoist room. For the hoists used in hot regions, cooling facilities may be provided in the hoist room.

4.0.4 The setting elevation of a hoist shall be such that the hoist and electrical equipment are prevented from being flooded and shall facilitate the maintenance of the gate, slot and hoist.

4.0.5 The lift of a hoist shall meet the operation and maintenance requirements of the gate, with a certain margin.

4.0.6 Hoists shall be so arranged as to prevent the wire ropes, movable pulleys, slings or tie rods from interfering with the gate slot and hydraulic structures.

4.0.7 When the hoist lifts a plane gate, its lifting centerline shall be consistent with that of the gate.

4.0.8 The movable pulleys shall be kept above the sediment accumulation elevation.

4.0.9 The hoist room shall be arranged away from the air vent for the gate.

4.0.10 An emergency power supply shall be provided for the hoist of the service gate for flood discharge, and a power-free emergency manipulator should be provided. A power-free emergency manipulator should be additionally provided for the diversion tunnel closure gate.

4.0.11 Hoists shall be able to be locally controlled. Gates with remote control requirements shall be provided with remote control interfaces and local video surveillance systems.

4.0.12 When a filling valve or a small opening is used for the pressure balance of the gate, the hoist shall be provided with a stroke detector for the small opening control.

4.0.13 The lowering speed of the hoist for a quick-acting shutoff gate at the power intake shall be determined according to the time required for closing the orifice, and a speed limiter shall be provided to control the lowering speed of the gate approaching the bottom sill within 5 m/min.

4.0.14 The closed gear transmission should be adopted for the hoist.

4.0.15 The hoist should be equipped with a safety brake.

4.0.16 The hoist frame shall have the required strength, stiffness and stability.

4.0.17 When the hoisting direction tilts, the horizontal load shall be considered in the calculation for the components of the hoist.

4.0.18 The hoist with the wire rope wound on the drum in multiple layers shall meet the following requirements:

1 The hoist with the wire rope wound on the drum in multiple layers should adopt the single lifting point type.

2 The hoist with the wire rope wound on the drum in three layers or more shall adopt a broken line groove drum.

3 For the hoist with two double-reeved pulley blocks with a ratio greater than 2, the fixed pulley shall be hinged to the pulley block support, and measures shall be taken to avoid the wire rope from interfering with the support beam of the fixed pulley.

4 Multi-layer winding wire ropes shall be pre-tensioned before delivery.

5 The lengths of the broken line section and straight section of a broken line groove drum shall be determined according to the diameters of the drum and wire rope.

6 A balance beam should be provided for the hoist with double lifting points and a lift greater than 100 m.

7 The hoist with the wire rope wound on the drum in multiple layers shall be provided with a flange at the return point of the wire rope.

8 When the hoist adopts a broken line groove drum, the number of layers for the wire rope wound on the drum should be no more than 5 for the hoist with a single lifting point, and no more than 3 for the hoist with double lifting points.

4.0.19 The hoists for radial gates shall meet the following requirements:

1 For the hoist with the lifting point arranged in front of the skin plate of an emersed radial gate, the wire rope shall be tightly against the skin plate and the slings shall not contact the skin plate.

2 For the hoist with double lifting points arranged behind the skin plate of an emersed radial gate, the winding of the wire rope and the arrangement of the pulley block shall meet the requirements for the synchronous operation of the lifting points.

3 For hoisting a submerged radial gate, the interference between wire ropes and the fixed pulley support beam shall be avoided. When the fixed pulley or the direction-guide pulley is installed beneath the support beam, access to maintenance and lubrication shall be available.

4 The incense coil hoist shall be provided with a wire rope length adjustment device. Wire ropes shall be pre-tensioned before delivery. The spacing between the wire rope catchers shall ensure that the wire rope is arranged neatly.

4.0.20 Hoists for tilting gates shall meet the following requirements:

1 The soffit of the hoist frame bridge shall be 0.3 m higher than the moving trajectory of the gate top.

2 During hoisting operation, the wire rope shall not rub against the gate leaf. After the gate is fully opened, the included angle between the line connecting the center of the lifting lug and the lifting center of the hoist and the vertical line shall not be greater than 15°.

4.0.21 For the hoist with double lifting points and the wire rope wound on the drum in multiple layers, the diameter deviation of the wire rope and the relative deviation between the tread diameters of two drums shall be controlled.

4.0.22 Measures shall be taken to ensure reliable and synchronous operation of the two lifting points of the hoist, and synchronization may be realized by a mechanical synchronizing shaft or by electrical means.

4.0.23 The movable pulleys shall be provided with a device or structure to prevent the wire rope from escaping from the groove. For the movable pulley with the rolling bearing flooded, a bearing sealing device shall be provided. For the movable pulley with the sliding bearing flooded, a bearing sealing device should be provided.

4.0.24 A heavy lifting shaft of the movable pulley(s) should be provided with a shaft shifting device with counterweight.

4.0.25 Lubricating oil and grease shall be selected according to the local environmental conditions and environmental protection requirements.

4.0.26 The anti-corrosion measures for hoists shall meet the following requirements:

1 Anti-corrosion materials or processes shall be selected for the hoist according to the working environment, environmental requirements, service life and operating conditions.

2 All exposed non-working surfaces and unenclosed cavities of the metal structural members of the hoist shall be subjected to corrosion treatment, and the corrosion control requirements shall comply with the current sector standard DL/T 5358, *Technical Code for Anticorrosion of Metal Structures in Hydroelectric and Hydraulic Engineering*.

3 The surfaces of the lifting shaft, pulley shaft and articulated shaft of the hoist shall be chrome plated for corrosion prevention.

4 Surfaces of submerged fasteners shall be galvanized for corrosion prevention or made of stainless steel.

5 Loads

5.0.1 The dead loads shall include the weights of the structural members, mechanical equipment and electrical equipment of the hoist.

5.0.2 The hoisting load shall be the maximum lifting force or the maximum holding force applied to the lifting lugs of the hoist connected to the gate or tie rod.

5.0.3 The temperature load need not be considered for the hoist.

5.0.4 The wind load and snow load need not be considered for the hoist itself, but shall be considered in the design of the hoist cover, and their values shall comply with the current national standard GB 50009, *Load Code for the Design of Building Structures*.

5.0.5 The seismic load shall be considered when the design seismic intensity is Ⅶ or above.

6 Materials

6.1 Castings

6.1.1 Carbon steel castings shall adopt ZG230-450, ZG270-500, ZG310-570, ZG340-640, or other materials stipulated in the current national standard GB/T 11352, *Carbon Steel Castings for General Engineering Purpose.*

6.1.2 Alloy steel castings shall adopt ZG35CrlMo, ZG42CrlMo, ZG40Crl, ZG65Mn, ZG40Mn2, ZG50Mn2, or other materials stipulated in the current sector standard JB/T 6402, *Heavy Low Alloy Steel Castings—Technical Specification.*

6.1.3 Grey iron castings shall adopt HT150, HT200, HT250, or other materials stipulated in the current national standard GB/T 9439, *Grey Iron Castings.*

6.1.4 Spheroidal graphite iron castings shall adopt QT450-10, QT500-7, or other materials stipulated in the current national standard GB/T 1348, *Spheroidal Graphite Iron Castings.*

6.1.5 Copper alloy castings shall adopt ZCuSn5Pb5Zn5, ZCuSn10P1, ZCuAl10Fe3, ZCuAl10Fe3Mn2, ZCuZn38Mn2Pb2, ZCuZn25Al6Fe3Mn3, or other materials stipulated in the current national standard GB/T 1176, *Cast Copper and Copper Alloys.*

6.2 Forgings and Rolled Pieces

6.2.1 Carbon steel forgings and rolled pieces shall adopt 20, 25, 35, 45, 50Mn, 65Mn, or other materials stipulated in the current national standards GB/T 699, *Quality Carbon Structural Steels;* and GB/T 33083, *Heavy Carbon Structural Steel Forgings—Technical Specification.*

6.2.2 Alloy steel forgings and rolled pieces shall adopt 35CrMo, 42CrMo, 40Cr, 40CrNi, 35SiMn, 42SiMn, 40MnB, or other materials stipulated in the current national standards GB/T 3077, *Alloy Structure Steels;* and GB/T 33084, *Heavy Alloy Structural Steel Forgings—Technical Specification.*

6.2.3 Stainless steel forgings and rolled pieces shall adopt materials stipulated in the current sector standard JB/T 6398, *Heavy Stainless Acid Resistant Steel and Heat Resistant Steel Forgings—Technical Specification.*

6.3 Structural Members

Load-bearing structural members shall adopt Q235 stipulated in the current national standard GB/T 700, *Carbon Structural Steels;* and Q355 stipulated in the current national standard GB/T 1591, *High Strength Low Alloy Structural*

Steels. They may also adopt Q390, Q420 and other materials stipulated in the current national standard GB/T 1591, *High Strength Low Alloy Structural Steels* when high strength steels are required. The material of load-bearing structural members shall be in accordance with Table 6.3.1.

Table 6.3.1 Materials of load-bearing structural members

Operating ambient temperature	Not lower than 0 °C	Not lower than –20 °C	Lower than –20 °C
Steel grade	Q235B, Q355B, Q390B, Q420B	Q235C, Q355C, Q390C, Q420C	Q235D, Q355D, Q390D, Q420D

NOTE The operating ambient temperature is determined based on the annual mean minimum daily temperature at the place where the hoist works.

6.4 Connecting Materials

6.4.1 Welding materials shall meet the following requirements:

1 Electrodes for arc welding shall adopt the relevant types stipulated in the current national standards GB/T 5117, *Covered Electrodes for Manual Metal Arc Welding of Non-alloy and Fine Grain Steels;* and GB/T 5118, *Covered Electrodes for Manual Metal Arc Welding of Creep-Resisting Steels*. The types of electrodes shall be compatible with the strength of the base metal.

2 Wire electrodes and fluxes for submerged arc welding shall comply with the current national standards GB/T 5293, *Solid Wire Electrodes, Tubular Cored Electrodes and Electrode/Flux Combinations for Submerged Arc Welding of Non Alloy and Fine Grain Steels*; and GB/T 12470, *Solid Wire Electrodes, Tubular Cored Electrodes and Electrode/Flux Combinations for Submerged Arc Welding of Creep-Resisting Steels*. The types of wire electrodes and fluxes shall be compatible with the strength of the base metal.

3 Wire electrodes for gas shielded welding shall comply with the current standards of China GB/T 8110, *Welding Electrodes and Rods for Gas Shielding Arc Welding of Carbon and Low Alloy Steel*; GB/T 10045, *Tubular Cored Electrodes for Non-alloy and Fine Grain Steels*; GB/T 14957, *Steel Wires for Melt Welding*; GB/T 17493, *Tubular Cored Electrodes for Creep-Resisting Steels*; GB/T 17853, *Tubular Cored Electrodes for Stainless Steels*; and YB/T 5092, *Stainless Steel Wires for Welding*.

6.4.2 The materials of fasteners shall meet the following requirements:

1 General purpose bolts, screws and studs shall be made of materials stipulated in the current national standards GB/T 3098.1, *Mechanical Properties of Fasteners —Bolts, Screws and Studs*; and GB/T 3098.3, *Mechanical Properties of Fasteners —Set Screws*. Plain nuts shall be made of materials stipulated in the current national standard GB/T 3098.2, *Mechanical Properties of Fasteners—Nuts*.

2 Stainless steel bolts, screws and studs shall be made of materials stipulated in the current national standard GB/T 3098.6, *Mechanical Properties of Fasteners—Stainless Steel Bolts, Screws and Studs*, and stainless steel nuts shall be made of materials stipulated in the current national standard GB/T 3098.15, *Mechanical Properties of Fasteners—Stainless Steel Nuts*.

3 High-strength bolts, nuts and washers shall be made of materials stipulated in the current national standards GB/T 3632, *Sets of Torshear Type High Strength Bolt Hexagon Nut and Plain Washer for Steel Structures*; and GB/T 1231, *Specifications of High Strength Bolts with Large Hexagon Head, Large Hexagon Nuts, Plain Washers for Steel Structures*.

6.4.3 The material of load-bearing connecting pins should be Grade 45 steel stipulated in the current national standard GB/T 699, *Quality Carbon Structural Steels*; and 35CrMo, 42CrMo and other materials stipulated in GB/T 3077, *Alloy Structure Steels*. The materials should be subjected to heat treatment as necessary.

7 Mechanical Design

7.1 Design and Calculation of Hoisting Mechanisms

7.1.1 The selection of motors shall meet the following requirements:

1 Motors shall be selected based on the following factors:

1) Type and structural form;

2) Power supply mode;

3) Power;

4) Duty and cyclic duration factor;

5) Rated torque, locked-rotor torque and maximum torque;

6) Rated speed;

7) Speed regulation mode;

8) IP degree;

9) Ambient temperature and humidity;

10) Altitude.

2 The selection of parameters and types for motors shall meet the following requirements:

1) Fixed wire rope hoists should adopt wound-rotor asynchronous motors, squirrel cage asynchronous motors, self-braking asynchronous motors, or AC variable-frequency motors.

2) The static power of motors shall be calculated by the rated hoisting load, sling weight, rated hoisting speed and mechanism efficiency, and motors shall be selected according to the static power, duty type, and cyclic duration factor or load duration. In this case, motors need not be subjected to overload and heating check.

3) The static power of motors may also be calculated by the equivalent hoisting load, sling weight, rated hoisting speed and mechanism efficiency. Motors may be selected according to the static power, operating mode, and cyclic duration factor or load duration. In this case, motors shall be checked for overload and heating in accordance with Appendix B of this code.

3 The torque on the motor shaft during steady lifting of the rated lifting load shall be calculated by the following formula:

$$M_N = \frac{(P_Q + q)D}{2ai\eta} \quad (7.1.1\text{-}1)$$

where

M_N is the torque on the motor shaft during steady lifting of the rated lifting load (N · m);

P_Q is the rated lifting load (N);

q is the weight of slings and wire ropes (N);

D is the pitch diameter, i.e. the distance from center to center of a rope wound on a drum (m);

a is the ratio for pulley block system;

i is the total ratio from the motor shaft to the drum shaft;

η is the overall efficiency of the transmission device and pulley block of the hoisting mechanism, and the approximate mechanical transmission efficiency for different parts and components shall be selected according to Table 7.1.1.

Table 7.1.1 Approximate mechanical transmission efficiency

<table>
<tr><th colspan="2" rowspan="2">Transmission parts and components</th><th colspan="2">Efficiency</th></tr>
<tr><th>Sliding bearing</th><th>Rolling bearing</th></tr>
<tr><td colspan="2">Open cylindrical gear pair (grease lubrication)</td><td rowspan="2">0.90 - 0.92</td><td>0.92 - 0.94</td></tr>
<tr><td colspan="2">Closed cylindrical gear pair (oil lubrication)</td><td>0.96 - 0.98</td></tr>
<tr><td colspan="2">Open bevel gear pair (grease lubrication)</td><td rowspan="2">0.90 - 0.92</td><td>0.92 - 0.94</td></tr>
<tr><td colspan="2">Closed bevel gear pair (oil lubrication)</td><td>0.95 - 0.97</td></tr>
<tr><td colspan="2">Intermediate shaft</td><td>0.95 - 0.97</td><td>0.97 - 0.99</td></tr>
<tr><td colspan="2">Drum</td><td>0.94 - 0.96</td><td>0.96 - 0.98</td></tr>
<tr><td colspan="2">Pulley</td><td>0.95</td><td>0.98</td></tr>
<tr><td colspan="2">Gear coupling</td><td colspan="2">0.99</td></tr>
<tr><td rowspan="2">Speed reducer</td><td>Moderately hard tooth-surface gear transmission at each stage</td><td colspan="2">0.98</td></tr>
<tr><td>Hard tooth-surface gear transmission at each stage</td><td colspan="2">0.98 - 0.985</td></tr>
</table>

Table 7.1.1 *(continued)*

Transmission parts and components		Efficiency		
	Ratio for pulley block system	Sliding bearing	Rolling bearing	Composite sliding and rolling bearings
Pulley block	2	0.97	0.99	0.97
	4	0.92	0.97	0.94
	6	0.88	0.95	0.91
	8	0.84	0.93	0.88
	10	0.80	0.91	0.85
	12	0.76	0.89	0.82

4 The torque on the motor shaft at the rated holding load shall be calculated by the following formula:

$$M_N = \frac{(P_c + q)D}{2ai} \tag{7.1.1-2}$$

where

M_N is the torque on the motor shaft at the rated holding load (N · m);

P_c is the rated holding load (N);

q is the weight of slings and wire ropes (N);

D is the pitch diameter (m);

a is the ratio for pulley block system;

i is the total ratio from the motor shaft to the drum shaft.

5 To accelerate lifting the rated load and compensate for the torque loss caused by changes in power voltage and frequency, the starting torque of motors shall meet the following requirements:

1) For line-start squirrel cage asynchronous motors, Formula (7.1.1-3) shall be satisfied:

$$M_d \geq 1.6M_N \tag{7.1.1-3}$$

where

M_d is the starting torque (N · m);

M_N is the torque on the motor shaft during steadily lifting or holding the rated load (N · m).

2) For wound-rotor asynchronous motors, Formula (7.1.1-4) shall be satisfied:

$$M_d \geq 1.9M_N \quad (7.1.1\text{-}4)$$

3) For motors controlled by variable frequency, Formula (7.1.1-5) shall be satisfied:

$$M_d \geq 1.4M_N \quad (7.1.1\text{-}5)$$

6 The degrees of protection of motors shall meet the following requirements:

1) The degree of protection of motor enclosure shall comply with the current national standard GB/T 4942.1, *Degrees of Protection Provided by the Integral Design of Rotating Electrical Machines (IP Code)—Classification.*

2) The degree of protection shall be IP44 or higher for indoor application.

3) The degree of protection shall be IP54 or higher for outdoor application. If condensate water is likely to present, the drain hole of condensate water shall be unblocked.

4) When the motor is protected by external means, a lower degree of protection may be adopted.

7 When the motor operates at an altitude exceeding 1,000 m or the operating ambient temperature is inconsistent with its rated ambient temperature, the power correction of the motor shall be in accordance with Appendix C of this code.

7.1.2 Brakes shall meet the following requirements:

1 Supporting brakes shall meet the following requirements:

1) The type selection of supporting brakes shall comply with the current sector standards JB/T 6406, *Electro-Hydraulic Drum Brakes*; JB/T 7685, *Electro-Magnetic Drum Brakes*; and JB/T 7020, *Electro-Hydraulic Disc Brakes.*

2) Supporting brakes shall be normally engaged, with the brake wheels or discs rigidly assembled on the high-speed shaft of the speed reducer. The disc brakes should be installed symmetrically; otherwise, relevant components shall be checked.

3) At least one supporting brake shall be provided for each independent drive. For important hoists, two supporting brakes shall be provided for each independent drive.

4) The braking torque of supporting brakes shall not be less than the calculated braking torque required for the brake shaft and calculated by the following formula:

$$M_Z = K_Z \frac{(P_Z + q)D_{\eta'}}{2ai} \tag{7.1.2-1}$$

where

M_Z is the calculated braking torque on the brake shaft (N · m);

K_Z is the braking safety factor;

P_Z is the maximum braking load during hoisting (N);

q is the weight of slings and wire ropes (N);

D is the pitch diameter (m);

η' is the overall efficiency of the transmission and pulley block of the hoisting mechanism when the gate is closing;

a is the ratio for pulley block system;

i is the total ratio between the brake shaft and the drum shaft.

5) For a drive with only one brake, the braking safety factor calculated by the total braking torque shall not be less than 1.75. For a drive with two brakes, the safety factor of each brake calculated by the total braking torque shall not be less than 1.25. For two drives in rigid connection, each with one brake, the safety factor of each brake calculated by the total braking torque shall not be less than 1.25. For two drives in rigid connection, each with two brakes, the safety factor of each brake calculated by the total braking torque shall not be less than 1.1. For planetary differential speed reducer transmission, each drive shall be provided with two supporting brakes, and the braking safety factor of each brake calculated by the total braking torque shall not be less than 1.75.

2 Safety brakes shall meet the following requirements:

1) The type selection of safety brakes may comply with the current sector standard JB/T 7020, *Electro-Hydraulic Disc Brakes*.

2) If the supporting brake is engaged under normal braking conditions,

the safety brake shall be engaged with a delay of 1 s to 2 s.

3) In the case of overspeed due to mechanism failure or transmission damage, the safety brakes shall be engaged automatically when the lowering speed of the gate reaches 1.5 times the specified speed.

4) When the hoist is operated, the safety brake is disengaged before the supporting brake is disengaged.

5) The braking torque of the safety brake shall not be less than the braking torque on the brake disc calculated by the following formula:

$$M_A = K_A \frac{(P_A + q)D}{2a} \tag{7.1.2-2}$$

where

M_A is the calculated braking torque on the brake disc (N · m);

K_A is the braking safety factor of the safety brake, taken as 1.50 to 1.75;

P_A is the maximum braking load of a single drum when hoisting the gate (N);

q is the weight of slings and wire ropes on a single drum (N);

D is the pitch diameter (m);

a is the ratio for pulley block system.

3 Retarding braking shall meet the following requirements:

1) Retarding braking is applied to reduce the lowering speed of a gate to zero or to a lower speed for braking.

2) Retarding braking of the hoisting mechanism may be realized by the supporting brake or by the electric braking.

3) Electric braking is only used for retarding braking, and shall be applied before mechanical braking.

7.1.3 The time and acceleration/deceleration for starting/braking the hoisting mechanism shall meet the following requirements:

1 The time and acceleration/deceleration for starting/braking the hoisting mechanism shall comply with Appendix D of this code, and the average acceleration/deceleration for starting/braking the hoisting mechanisms shall be less than 0.3 m/s^2.

2 In the case of AC variable frequency speed regulation and electric braking, the acceleration/deceleration for the hoisting mechanism need not be calculated.

7.1.4 The selection of speed reducers shall meet the following requirements:

1 An appropriate reducer shall be selected according to the calculated load and total ratio of the hoisting mechanism.

2 The reducer of the hoist should adopt a closed gear transmission directly connecting the drum.

3 The reducer of the hoist may consist of a standard crane speed reducer and an open gear transmission, and the single-stage transmission ratio of the open gear should not exceed 6.3.

7.2 Calculation Principles for Parts and Components

7.2.1 The calculation of parts and components shall meet the following requirements:

1 The strength calculation shall include the static strength and fatigue strength.

2 The strength calculation may adopt the allowable stress method and the safety factor method.

3 Stiffness and stability shall be calculated for the parts and components such as hoist drums.

4 The critical speed of long high-speed transmission shafts shall be checked.

5 The calculation of parts and components shall comply with Appendix E of this code.

7.2.2 The calculation loads of parts and components shall meet the following requirements:

1 The maximum service load shall be used to calculate the static strength of the parts and components. For the hoisting mechanism, the maximum service load is taken as 1.1 to1.2 times the torque or force transmitted to the parts and components from the hoisting force; for the parts and components on the high-speed shaft, the maximum service load is taken as 2.0 to 2.5 times the rated torque of the motor.

2 The basic load shall be used to calculate the fatigue strength of the parts and components. For the parts and components of the hoisting mechanism, the basic load is taken as 0.6 to 1.0 times the torque or force transmitted to the parts and components from the hoisting force

according to the gate type and working properties; for the parts and components on the high-speed shaft, the basic load is taken as 1.3 to 1.4 times the rated torque of the motor.

7.2.3 The yield strength limits of parts and components shall meet the following requirements:

1 In the calculation of static strength of parts and components, when the ratio of the lower yield strength R_{eL} to the tensile strength R_m of the material is less than 0.7, the yield strength limit of the material shall be used as the yield strength of parts and components.

2 When the ratio of the lower yield strength R_{eL} to the tensile strength R_m of the material is 0.7 or larger, the hypothetical yield strength shall be calculated by the following formulae:

$$R_{sF}=\frac{R_{eL}+0.7R_m}{2} \tag{7.2.3-1}$$

$$\tau_{sF}=\frac{R_{sF}}{\sqrt{3}} \tag{7.2.3-2}$$

where

R_{sF} is the calculated hypothetical yield strength (N/mm^2);

R_{eL} is the lower yield strength of material (N/mm^2);

R_m is the tensile strength of material (N/mm^2);

τ_{sF} is the calculated hypothetical shear strength (N/mm^2).

7.2.4 The strength of parts and components shall be checked according to the following requirements:

1 The strength of mechanical transmission parts and components shall be checked; however, for the parts and components subjected to load infrequently or for a short time, the fatigue strength need not be checked since no fatigue damage would occur.

2 In strength checking, the calculated stress of the parts and components shall not exceed the value obtained by dividing the yield strength-limit or fatigue strength-limit of the material of the parts and components by the strength safety factor.

3 In strength checking, the strength safety factor shall be determined with comprehensive consideration of the factors such as load, mechanism properties and material process.

7.2.5 The calculation of fatigue strength shall comply with the current national standard GB/T 3811, *Design Rules for Cranes*.

7.2.6 For the parts and components worn frequently during operation, the pressure per unit area of the section, p, and pv, the product of p and the relative motion velocity v of the faying surface, shall be checked to make sure that the values are not greater than their respective allowable values.

7.3 Design of Parts and Components

7.3.1 Shafts shall meet the following requirements:

1 Shafts shall be made of materials, such as 35 and 45 and other carbon structural steels, stipulated in the current national standard GB/T 699, *Quality Carbon Structural Steels*; and 35CrMo, 42CrMo, 40Cr, 40CrNi, 35SiMn, 42SiMn, 40MnB and other alloy structural steels stipulated in the current national standard GB/T 3077, *Alloy Structure Steels*.

2 In the most unfavorable case, the calculation load of the shaft shall be the maximum service load, and the shaft may be regarded as operating under static load. The calculation of shafts shall comply with Section E.1 of this code.

3 When the speed of the drive shaft is greater than 400 r/min, in addition to calculating the strength and stiffness, the critical speed shall be checked and satisfy the following formulae:

$$n_{max} \leq \frac{n_{cr}}{1.2} \tag{7.3.1-1}$$

$$n_{cr} = 121\frac{\sqrt{d_1^2 + d_2^2}}{L^2} \tag{7.3.1-2}$$

where

n_{max} is the actual maximum speed of shaft (r/min);

n_{cr} is the critical speed (r/min);

d_1 is the inner diameter of shaft (mm), taken as 0 for a solid shaft;

d_2 is the outer diameter of shaft (mm);

L is the distance between fulcrums of the shaft (m).

4 The calculation of shaft stiffness should meet the following requirements:

1) The maximum deflection should not exceed 1/3000 of the distance between fulcrums.

2) The maximum deflection of the shaft with gear should not exceed 0.03 times the gear modulus.

3) The maximum deflection angle caused by deflection at the fulcrum should not exceed 0.001rad.

4) The allowable torsion angle should not exceed 0.5°/m.

5 For the hoists with double lifting points and a mechanically synchronizing shaft, the shaft shall be capable of transmitting the torque required for one lifting point.

7.3.2 The calculation of lifting pallets should comply with Section E.2 of this code. Materials selected shall be of mechanical properties not inferior to Q235 in the current national standard GB/T 700, *Carbon Structural Steels*; or not inferior to Q355 in the current national standard GB/T 1591, *High Strength Low Alloy Structural Steels*. Meanwhile, the materials shall be in accordance with Table 6.3.1.

7.3.3 Wire ropes shall meet the following requirements:

1 Wire ropes for hoists shall comply with the current standards of China GB/T 8918, *Steel Wire Ropes for Important Purposes*; and YB/T 5359, *Compacted Strand Rope*.

2 Galvanized wire ropes should be used. In the case of frequently operating in water, galvanized wire ropes shall be used.

3 Single-layer winding wire ropes should be of synthetic fiber core, and multi-layer winding wire ropes shall be of metal core.

4 The selection and calculation of wire ropes shall meet the following requirements:

1) The maximum static tension of wire ropes in operation shall be calculated by the following formulae:

$$S=\frac{P+q}{xa\eta_{\Sigma}} \tag{7.3.3-1}$$

$$\eta_{\Sigma}=\frac{1-\eta_1^a}{(1-\eta_1)a}\eta_{\mathrm{D}} \tag{7.3.3-2}$$

where

S is the maximum static tension of wire rope in operation (N);

P is the rated lifting force or rated holding force (N);

q is the weight of slings and wire ropes (N);

x is the number of parts of wire rope wound on the drum, taken as 2.0 for double reeving and 1.0 for single reeving;

a is the ratio for pulley block system;

η_{Σ} is the overall transmission efficiency of wire rope system;

η_1 is the efficiency of a single pulley, which is selected as per Table 7.1.1 ;

η_D is the efficiency of the guide pulley, which is selected as per Table 7.1.1 and is taken as 1 in the case of no guide pulley.

2) Wire ropes shall be selected according to the safety factor related to the duty class of hoists, and the minimum breaking force of the entire wire rope selected shall satisfy the following formula:

$$F_0 \geq nS \tag{7.3.3-3}$$

where

F_0 is the minimum breaking force of the entire wire rope (N);

n is the safety factor of wire rope, which is selected according to Table 7.3.3 For the wire rope operating in water over a long period of time, the safety factor may be appropriately increased according to the local water quality conditions;

S is the maximum static tension of wire rope in operation (N).

Table 7.3.3 Safety factor *n* of wire rope

Duty class of hoist	Safety factor *n*
Q1	4.5
Q2	5.0
Q3, Q4	5.5

7.3.4 Pulleys and drums shall meet the following requirements:

1 The tread diameters of pulleys and drums shall meet the following requirements:

1) The tread diameter of the pulley or drum shall be calculated by the following formula according to the nominal diameter of the wire rope:

$$D = ed \tag{7.3.4-1}$$

where

D is the tread diameter of pulley or drum (mm);

e is the ratio of the tread diameter of drum or pulley to the diameter of wire rope, which shall not be less than the value given in Table 7.3.4;

d is the nominal diameter of wire rope (mm).

2) The tread diameter of the equalizer pulley should be $0.6D$ to $0.8D$.

Table 7.3.4 Ratio of tread diameter of drum or pulley to diameter of wire rope

Duty class of hoist	Ratio *e*
Q1, Q2	18
Q3	20
Q4	22

2 The materials of pulleys and drums shall meet the following requirements:

1) Cast pulleys and drums shall be made of materials with mechanical properties not inferior to HT200 in the current national standard GB/T 9439, *Grey Iron Castings*; and ZG270-500 in the current national standard GB/T 11352, *Carbon Steel Castings for General Engineering Purpose*.

2) Welded and rolled pulleys and drums shall be made of materials with mechanical properties not inferior to Q235 in the current national standard GB/T 700, *Carbon Structural Steels*; or not inferior to Q355 in the current national standard GB/T 1591, *High Strength Low Alloy Structural Steels*. Meanwhile, they shall be in accordance with Table 6.3.1.

3) For large hoists, the pulleys should be made of cast steel or made by welding and rolling, and the drums should be made of steel plates by rolling and welding or made of cast steels.

3 The structural types of pulleys and drums should meet the following requirements:

1) The structural type of the pulley should comply with the current national standard GB/T 27546, *Sheaves for Cranes*.

2) The structural type of the drum shall comply with the current sector standard JB/T 9006, *Drums for Cranes*.

3) When the welded drum is of the short shaft type, it may be connected to the speed reducer via a drum coupling.

4 The strength calculation and stability check for drums shall meet the following requirements:

1) When the length of the drum is less than or equal to 3 times the tread diameter of the drum, only the maximum compressive stress on the wall surface of the drum is required to be calculated.

2) When the length of the drum is greater than 3 times the tread diameter of the drum, in addition to the compressive stress calculation, the resultant stress generated by bending moment and torque shall be checked.

3) When the tread diameter of the drum is not less than 1200 mm or the length of the drum is greater than 2 times the tread diameter of the drum, in addition to the strength calculation, the stability of the drum wall shall be checked.

4) The calculation of drums shall comply with Section E.3 of this code.

5 The fleet angle of the wire rope shall meet the following requirements:

1) When the wire rope is wound in or out of the pulley groove, the maximum fleet angle of the wire rope shall not be greater than 5°.

2) When the wire rope is wound in or out of the drum, the fleet angle of the wire rope centerline from the groove centerline shall not be greater than 3.5° on either side.

3) For the double-layer winding spiral groove drum with a single lifting point, the fleet angle of the wire rope at the return point from the vertical plane of the drum shaft shall not be greater than 2°, and should not be less than 0.5°.

4) For the double-layer winding spiral groove drum with double lifting points, the fleet angle of the wire rope at the return point from the vertical plane of the drum shaft shall not be greater than 1.5°, and should not be less than 0.5°.

5) For the multi-layer winding broken line groove drum, the fleet angle of the wire rope at the return point from the vertical plane of the drum shaft shall not be greater than 1.5°, and should not be less than 0.5°.

6 When the wire rope is fixed by clamping plate bolts, the tension at the position where the wire rope is fixed, and the pressing force and tensile

stress of the clamping plate bolt shall be calculated in accordance with Section E.3 of this code. Clamping plates for fixing round wire ropes shall be selected in accordance with the current national standard GB/T 5975, *Clamping Plates for Fixing Steel Wire Ropes*.

7.3.5 Gear transmission shall meet the following requirements:

1 Pinions should be made of quality carbon steel or alloy structural steel, and large gears should be made of cast carbon steel or cast alloy steel. The requirements of gear matching shall be considered in selection of material and heat treatment hardness.

2 In the case of soft or moderately hard tooth-surface gear pairs, the tooth surface hardness of pinions shall be greater than that of large gears, and the difference shall be taken between 30 HB and 50 HB.

3 The tooth surface contact strength and gear bending strength shall be calculated for gear transmission. Closed gear transmission should have a medium or hard tooth surface; the load capacity of involute cylindrical gears shall be calculated in accordance with the current national standard GB/T 3480, *Calculation Methods of Load Capacity for Involute Cylindrical Gears*.

7.3.6 Speed reducers shall meet the following requirements:

1 When a standard speed reducer is selected, the duty class shall be compatible with that of the hoist. Speed reducers shall be selected based on the rated load or motor rated power and the required operating conditions, and the maximum radial load at the output shaft end of the speed reducer shall be checked if necessary.

2 The heavy-duty reducer box should be of the welded or cast type, and the gear should have a moderately hard or hard tooth surface.

3 Oil bath lubrication should be adopted for speed reducers. Spray lubrication shall be adopted for heavy-duty reducers and shall be performed for high-speed mesh gears before startup.

7.3.7 Couplings shall meet the following requirements:

1 The type of couplings for hoists may be determined according to the operating conditions, and then the coupling may be selected from the table of standard specifications of couplings according to the calculation torque of the coupling and the journal size and speed of the connected shaft.

2 The torque of couplings shall be calculated by Formula (7.3.7-1) and shall satisfy Formula (7.3.7-2):

$$M_{L}=kM'_{L} \tag{7.3.7-1}$$

$$M_{L}\leq[M_{L}] \tag{7.3.7-2}$$

where

M_{L} is the calculated torque of coupling (N · m);

k is the safety factor, which is taken as 2.5 for the coupling at the high-speed end of the speed reducer, 1.5 for the synchronous shaft coupling, and no less than 1.25 for the coupling connecting the drum and the low-speed shaft of the speed reducer;

M'_{L} is the torque transmitted from the coupled shaft, which is calculated by the rated torque of the motor (N · m);

$[M_{L}]$ is the allowable torque of coupling (N · m).

7.3.8 Bearings shall meet the following requirements:

1 Sliding bearings generally used for low-speed and heavy-duty transmission of hoists shall be of the self-lubricating type. The allowable values of common sliding bearing materials shall comply with Appendix F of this code.

2 Equivalent dynamic and static loads shall be calculated based on the following conditions, then the rated dynamic and static loads required shall be calculated, and the rolling bearing shall be selected:

1) The entire design service life of the bearing should be consistent with the duty class of the hoist.

2) For rolling bearings with a speed less than 10 r/min, only the rated static load is calculated.

3) Imposed radial load.

4) Imposed axial load.

5) Operating conditions, nature of loads, rotating race, type and supply method of lubricating oil.

6) Structural type and overall dimensions of bearings.

7) In the basic rated static load calculation of rolling bearings, the safety factor shall not be less than 2.0 for ball bearings, 3.0 for roller bearings, and 4.0 for thrust self-aligning roller bearings.

7.3.9 Load limiters shall meet the following requirements:

1 Each set of hoisting mechanism shall be provided with a load limiter consisting of load sensors, secondary instruments and corresponding accessories, with a combined error not exceeding 5 %. An audible and visual alarm signal shall be given when the lifting load reaches 95 % of the rated load of the hoist. During gate hoisting, when the hoisting load is between 100 % and 110 % of the rated load, an overload alarm signal shall be given and the power supply will be automatically switched off. There is no limit to overload during the closing of quick-acting shutoff gates or emergency gates.

2 Underload protection should be provided for hoists.

3 The calibrated values of lifting force and holding force of gates shall be adjusted according to the hoisting needs of gates, and the lifting force and holding force may be limited separately if necessary.

7.3.10 Height indicators and position limiters shall meet the following requirements:

1 Height indicators should be of the electric type and absolute sensors should be selected.

2 Hoists shall be equipped with a height indicator as well as electric and mechanical position limiters that can both be involved in control of the upper and lower limit positions. When the hoist reaches a position limit, the position limiters shall send out signal and switch off the power at the same time, but reverse operation shall be allowed.

3 When the hoist is provided with electrically synchronized double lifting points, a height limiter and a control device shall be provided for each lifting point, and the detection accuracy shall meet the requirements of synchronization.

4 For hoists operating the gate that balances pressure by filling water via a small opening or the gate with a filling valve, the height indicators and control devices shall be able to control positions in a certain range. When the height indicator shows that the control position setpoint is reached, the power supply shall be automatically switched off, but reverse operation shall be allowed.

8 Structural Design

8.1 Calculation Principles

8.1.1 The allowable stress method is used for structural calculation. The strength, stiffness and stability shall be calculated, and the effect of material plasticity need not be considered. The fatigue strength need not be calculated.

8.1.2 The structural calculation shall be conducted for two load combination types. For Type Ⅰ load combination, the strength, stiffness and stability shall be calculated assuming that the structure bears the maximum working load; for Type Ⅱ load combination, the strength and stability shall be checked assuming that the structure bears the maximum non-working load or the special working load.

8.2 Load Combinations

The load combinations used for calculation shall be in accordance with Table 8.2.1 and the installation loads may be added when necessary.

Table 8.2.1 Load combinations

Load	Type Ⅰ	Type Ⅱ	
		$Ⅱ_a$	$Ⅱ_b$
Dead load	√	√	√
Hoisting load	√	–	–
Weight of gate	–	–	√
Seismic load	–	√	√

8.3 Allowable Stresses

8.3.1 The allowable stresses of structural materials shall meet the following requirements:

1 The allowable tensile, compressive and bending stresses of structural materials are calculated as follows:

1) When the yield ratio R_{eH}/R_m of steel is less than 0.7, the allowable stress shall be calculated by the following formula:

$$[\sigma] = R_{eH} / n \tag{8.3.1-1}$$

where

$[\sigma]$ is the allowable stress of steel (N/mm^2);

R_{eH} is the upper yield strength of steel (N/mm^2);

R_m is the tensile strength of steel (N/mm^2);

n is the strength safety factor related to the load combination type, which shall be in accordance with Table 8.3.1-1.

Table 8.3.1-1 Strength safety factor *n*

Load combination	Type I	Type II
Safety factor *n*	1.48	1.22

2) When the yield ratio R_{eH}/R_m of steel is 0.7 or larger, the allowable stress shall be calculated by the following formula:

$$[\sigma]=\frac{0.5R_{eH}+0.35R_m}{n} \tag{8.3.1-2}$$

where

R_{eH} is the upper yield strength of steel (N/mm^2);

R_m is the tensile strength of steel (N/mm^2);

n is the strength safety factor related to the load combination type, which shall be in accordance with Table 8.3.1-1.

2 The allowable shear stress shall be calculated by the following formula:

$$[\tau]=\frac{[\sigma]}{\sqrt{3}} \tag{8.3.1-3}$$

where

$[\tau]$ is the allowable shear stress of steel (N/mm^2);

$[\sigma]$ is the allowable stress of steel (N/mm^2).

3 The allowable compressive stress of end face shall be calculated by the following formula:

$$[\sigma_{cd}] = 1.4\,[\sigma] \tag{8.3.1-4}$$

where

$[\sigma_{cd}]$ is the allowable compressive stress of end face (N/mm^2);

$[\sigma]$ is the allowable stress of steel (N/mm^2).

4 The allowable local contact bearing stress shall be calculated by the following formula:

$$[\sigma_{cj}] = 0.75\,[\sigma] \tag{8.3.1-5}$$

where

$[\sigma_{cj}]$ is the allowable local contact bearing stress (N/mm^2);

$[\sigma]$ is the allowable stress of steel (N/mm^2).

5 The size grouping and yield strength of common structural materials shall be in accordance with Table 8.3.1-2, and the allowable stresses of structural materials commonly used for Type Ⅰ load combination shall be in accordance with Table 8.3.1-3.

Table 8.3.1-2 Size grouping and yield strength of common structural materials

Group	Q235		Q355	
	Steel thickness (mm)	R_{eH} (N/mm^2)	Steel thickness (mm)	R_{eH} (N/mm^2)
Group 1	≤ 16	235	≤ 16	355
Group 2	> 16 to 40	225	> 16 to 40	345
Group 3	> 40 to 60	215	> 40 to 63	335
Group 4	> 60 to 100	215	> 63 to 80	325

Table 8.3.1-3 Allowable stresses of structural materials commonly used for Type I load combination (N/mm^2)

Stress type	Symbol	Q235				Q355			
		Group 1	Group 2	Group 3	Group 4	Group 1	Group 2	Group 3	Group 4
Tensile, compressive or bending stress	$[\sigma]$	160	150	145	145	240	235	225	220
Shear stress	$[\tau]$	95	90	85	85	140	135	130	125
Local compressive stress (tight fit)	$[\sigma_{cd}]$	225	210	200	200	335	325	315	305
Local contact bearing stress	$[\sigma_{cj}]$	120	115	110	110	180	175	170	165

NOTES:

1 Local compression happens when a small surface or end face of the web of a member is subjected to compression.

2 Local contact bearing stress refers to the compressive stress on the projection plane of the contact surface with a less mobility.

8.3.2 The allowable stress of welds shall meet the following requirements:

1 In the design of welded connection, the weld shall have the same comprehensive mechanical properties as the base metal.

2 According to the welding conditions, welding methods and weld types, the allowable stresses of welds shall be in accordance with Table 8.3.2-1, and the allowable stresses of welds used for structural materials commonly used for Type Ⅰ load combination shall be in accordance with Table 8.3.2-2.

Table 8.3.2-1 Allowable stresses of welds (N/mm^2)

Weld type	**Class of weld**	**Allowable transverse tensile or compressive stress** $[\sigma_h]$	**Allowable shear stress** $[\tau_h]$
Butt weld	Class Ⅰ and Class Ⅱ	$[\sigma]$	$[\sigma]/\sqrt{3}$
	Class Ⅲ	$0.8[\sigma]$	$0.8[\sigma]/\sqrt{3}$
Fillet weld	Automatic and manual welding	–	$[\sigma 3]/\sqrt{2}$

NOTES:

1 The classification of welds shall comply with the current sector standard NB/T 35051, *Code for Manufacture Erection and Acceptance of Gate Hoists in Hydropower Projects*.

2 $[\sigma]$ is the allowable stress of the base metal.

3 For welds with poor construction conditions or welds bearing transverse loads, their allowable stresses should be appropriately reduced.

Table 8.3.2-2 Allowable stresses of welds of structural materials commonly used for Type Ⅰ load combination (N/mm^2)

Weld type	**Stress type**		**Symbol**	**Q235**				**Q355**			
				Group 1	**Group 2**	**Group 3**	**Group 4**	**Group 1**	**Group 2**	**Group 3**	**Group 4**
Butt weld	Compressive and tensile stresses	Class Ⅰ and Class Ⅱ	$[\sigma_h]$	160	150	145	145	240	235	225	220
		Class Ⅲ	$[\sigma_h]$	130	120	115	115	190	185	180	175
	Shear stress	Class Ⅰ and Class Ⅱ	$[\tau_h]$	95	90	85	85	140	135	130	125
		Class Ⅲ	$[\tau_h]$	80	75	70	70	110	110	105	100

Table 8.3.2-2 (*continued*)

Weld type	Stress type	Symbol	Q235				Q355			
			Group 1	Group 2	Group 3	Group 4	Group 1	Group 2	Group 3	Group 4
Fillet weld	Compressive, tensile and shear stresses	$[\tau_h]$	115	105	100	100	170	165	160	155

NOTES:

1 The classification of welds shall comply with the current sector standard NB/T 35051, *Code for Manufacture Erection and Acceptance of Gate Hoists in Hydropower Projects*.

2 The allowable stresses of overhead welds are obtained by multiplying the values in Table 8.3.2-2 by 0.8.

3 The allowable stresses of erection welds are obtained by multiplying the values in Table 8.3.2-2 by 0.9.

4 When a single-angle steel weldment is connected with one leg, for the equal angle steel, either leg may be used; for the unequal angle steel, the shorter leg shall be used, the allowable stresses of the connecting welds are obtained by multiplying the values in Table 8.3.2-2 by 0.85.

8.3.3 The allowable stresses of bolt and pin connections shall meet the following requirements:

1 The allowable stresses of bolts shall be selected by Grade A, B or C bolt connections.

2 The allowable stresses of bolt and pin connections shall be in accordance with Table 8.3.3-1, and the allowable stresses for bolt and pin connections made of structural materials commonly used for Type Ⅰ load combination shall be in accordance with Table 8.3.3-2.

Table 8.3.3-1 Allowable stresses of bolt and pin connections (N/mm^2)

Connection type	Stress type	Allowable stress of bolt and pin	Allowable stress of connected member
Grades A and B bolt connections (Type I holes)	Tensile stress	$0.8S_p/n$	–
	Shear stress	$0.6S_p/n$	–
	Compressive stress	–	$1.8[\sigma]$
Grade C bolt connection	Tensile stress	$0.8S_p/n$	–
	Shear stress	$0.6S_p/n$	–
	Compressive stress	–	$1.4[\sigma]$

Table 8.3.3-1 *(continued)*

Connection type	Stress type	Allowable stress of bolt and pin	Allowable stress of connected member
Pin connection	Bending	[σ]	–
	Shear stress	0.6[σ]	–
	Compressive stress	–	1.4[σ]

NOTES:

1 S_p is the guaranteed stress of the bolt corresponding to the performance level of the bolt, which is be selected in accordance with the current national standard GB/T 3098.1, *Mechanical Properties of Fasteners —Bolts, Screws and Studs*.

2 n is the safety factor, taken as per Table 8.3.1-1 of this code.

3 [σ] is the allowable stress of the steels corresponding to bolts, pin shafts, or members, taken as per Table 8.3.1-3 of this code.

4 The allowable compressive stress of pin should be appropriately reduced when the pin moves slightly during operation.

5 The following holes shall be taken as Type I: the hole drilled in an assembly to the design aperture; the hole drilled in an individual part or member with a drill die to the design aperture; the hole drilled or punched to a small aperture in an individual part and then counterbored in the assembly to the design aperture.

Table 8.3.3-2 Allowable stresses for bolt and pin connections made of structural materials commonly used for Type I load combination (N/mm^2)

Connection type	Stress type	Symbol	Performance level of bolt				Steel grade of member							
							Q235				Q355			
			L4.6	L4.8	L5.6	L8.8	Group 1	Group 2	Group 3	Group 4	Group 1	Group 2	Group 3	Group 4
Grades A and B bolt connections (Type I holes)	Tensile stress	[σ]	–	–	155	315	–	–	–	–	–	–	–	–
	Shear stress	[τ]	–	–	115	235	–	–	–	–	–	–	–	–
	Compressive stress	[σ_c]	–	–	–	–	290	270	260	260	430	420	405	395
Grade C bolt connection	Tensile stress	[σ]	125	165	–	–	–	–	–	–	–	–	–	–
	Shear stress	[τ]	95	125	–	–	–	–	–	–	–	–	–	–
	Compressive stress	[σ_c]	–	–	–	–	225	210	205	205	335	325	315	305
Pin connection (steel grade 35)	Bending stress	[σ]	135				–	–	–	–	–	–	–	–
	Shear stress	[τ]	80				–	–	–	–	–	–	–	–
	Compressive stress	[σ_c]	–				225	210	205	200	335	325	315	305

8.3.4 The allowable compressive stress of phase Ⅰ and phase Ⅱ concrete for embedded parts shall be in accordance with Table 8.3.4.

Table 8.3.4 Allowable compressive stress of concrete (N/mm^2)

Strength class of concrete	C15	C20	C25	C30	C40
Allowable compressive stress	5	7	9	11	14

8.4 Strength Calculation for Structural Members and Connections

8.4.1 The calculation of structural members shall meet the following requirements:

1 When the structural member is in tension, compression, bending or torsion, its strength may be calculated according to the general strength calculation formula, and the calculated stress shall be less than the allowable stress.

2 When the point load is applied on the upper flange of the beam, the local compressive stress of the web shall be calculated by Formula (8.4.1-1) and shall satisfy Formula (8.4.1-2):

$$\sigma_{\mathrm{m}} = \frac{p}{\delta\left(a + 2h_{\mathrm{y}}\right)} \tag{8.4.1-1}$$

$$\sigma_{\mathrm{m}} \leq [\sigma] \tag{8.4.1-2}$$

where

σ_{m} is the local compressive stress (N/mm^2);

p is the point load (N);

δ is the thickness of web (mm);

a is the length of point load (mm);

h_{y} is the distance from the top surface of the member to the upper edge of the calculation height of the web (mm);

$[\sigma]$ is the allowable stress of steel (N/mm^2).

3 The calculation of combined stress shall meet the following requirements:

1) When the normal stresses σ_{x} and σ_{y} and shear stress τ_{xy} are applied on the same calculation point of a member, the combined stress σ_d at this point shall be calculated by Formula (8.4.1-3) and shall satisfy Formula (8.4.1-4):

$$\sigma_d = \sqrt{\sigma_x^2 + \sigma_y^2 - \sigma_x\sigma_y + 3\tau_{xy}^2} \quad (8.4.1\text{-}3)$$

$$\sigma_d \leq 1.1[\sigma] \quad (8.4.1\text{-}4)$$

where

σ_d is the combined stress (N/mm²);

σ_x, σ_y are the normal stresses applied on the same calculation point of a member, with their respective positive or negative sign (N/mm²), either of which shall be less than the allowable stress $[\sigma]$ in Table 8.3.1-3 of this code;

τ_{xy} is the shear stress on the same calculation point of a member (N/mm²), which shall be less than the allowable shear stress $[\tau]$ in Table 8.3.1-3 of this code;

$[\sigma]$ is the allowable stress of steel (N/mm²).

2) When the normal stress σ, shear stress τ and local compressive stress σ_m are applied on the same calculation point of a member, the combined stress σ_d on this point shall be calculated by Formula (8.4.1-5) and shall satisfy Formula (8.4.1-6):

$$\sigma_d = \sqrt{\sigma^2 + \sigma_m^2 - \sigma\sigma_m + 3\tau^2} \quad (8.4.1\text{-}5)$$

$$\sigma_d \leq 1.1[\sigma] \quad (8.4.1\text{-}6)$$

where

σ_d is the combined stress (N/mm²);

σ is the normal stress (N/mm²), which is positive or negative;

σ_m is the local compressive stress (N/mm²), which is positive or negative;

τ is the shear stress (N/mm²);

$[\sigma]$ is the allowable stress of steel (N/mm²).

3) When only the tensile or compressive stress σ and shear stress τ are applied, the combined stress σ_d shall be calculated by Formula (8.4.1-7) and shall satisfy Formula (8.4.1-8):

$$\sigma_d = \sqrt{\sigma^2 + 3\tau^2} \tag{8.4.1-7}$$

$$\sigma_d \leq 1.1[\sigma] \tag{8.4.1-8}$$

where

σ_d is the combined stress (N/mm^2);

σ is the tensile or compressive stress (N/mm^2);

τ is the shear stress (N/mm^2);

$[\sigma]$ is the allowable stress of steel (N/mm^2).

8.4.2 The strength of axially tensioned and compressed structural members connected by friction-type high-strength bolts shall be calculated by Formula (8.4.2-1) and Formula (8.4.2-2), and shall satisfy Formula (8.4.2-3):

$$\sigma = \frac{N'}{A_j} \tag{8.4.2-1}$$

$$N' = N\left(1 - \mu\frac{Z_1}{Z}\right) \tag{8.4.2-2}$$

$$\sigma \leq [\sigma] \tag{8.4.2-3}$$

where

σ is the tensile or compressive stress of the members connected by high-strength bolts (N/mm^2);

N' is the axial force for checking the members connected by high-strength bolts (N);

A_j is the net section area of the checked structural member (mm^2);

N is the axial force applied on the connection (N);

μ is the mean slip coefficient of faying surface, which is taken as per Table 8.4.2;

Z_1 is the number of the outermost high-strength bolts on the side corresponding to the calculation section connection joint of the structural member;

Z is the total number of high-strength bolts on the structural member on the side corresponding to the joint connected at nodes or splices;

$[\sigma]$ is the allowable stress of steel (N/mm^2).

Table 8.4.2 Mean slip coefficient μ of faying surface

Treatment of contact surfaces of members at the joint	Steel grade of member	
	Q235	Q355 or higher
Sandblasting with hard quartz sand or cast steel grit	0.45	0.45
Shot blasting (sandblasting)	0.40	0.40
Derusting with a wire brush or untreated clean rolled surface	0.30	0.35
Applying inorganic zinc-rich paint after sandblasting	0.35	0.40

NOTES:

1 The derusting by wire brush shall proceed in the direction normal to the force direction.

2 When the connected members are of different steel grades, μ is taken corresponding to the member of the lower strength.

3 When applying inorganic zinc-rich paint after sandblasting, the surface roughness of steel is required to reach Sa2.5, and the coating thickness is 60 μm to 80 μm.

4 When other methods are adopted for treatment, the treatment process and mean slip coefficient shall be determined by test.

8.4.3 The strength calculation of welded connections shall meet the following requirements:

1 The stress of butt weld is calculated by the minimum plate thickness t in the connection. When it is impossible to weld with a run-off plate, the calculation length of each weld is taken as its actual length minus $2t$. The calculation of butt welds shall be made as follows:

1) For the butt weld subject to axial tension or compression, the calculated transverse tensile and compressive stresses shall be lower than the allowable values shown in Table 8.3.2-1 of this code.

2) For the butt weld subject to both bending moment and shear stress, the maximum normal stress at the dangerous point shall not be greater than the allowable transverse tensile and compressive stresses of welds shown in Table 8.3.2-1 of this code, and the maximum shear stress shall not be greater than the allowable shear stress of welds shown in Table 8.3.2-1 of this code.

3) For the butt weld subject to combined stress, the combined stress on the position where both normal stress and shear stress are larger shall be calculated by Formula (8.4.3-1), and shall satisfy Formula (8.4.3-2):

$$\sigma_h = \sqrt{\sigma^2 + 3\tau^2} \quad (8.4.3\text{-}1)$$

$$\sigma_h \leq 1.1[\sigma_h] \quad (8.4.3\text{-}2)$$

where

σ_h is the combined stress of butt weld (N/mm^2);

σ is the tensile or compressive stress (N/mm^2);

τ is the shear stress (N/mm^2);

$[\sigma_h]$ is the allowable stress of weld (N/mm^2), which shall be taken as per Table 8.3.2-2 of this code.

2 The calculation of fillet welds shall meet the following requirements:

1) Under the action of tensile, compressive and shear forces through the centroid of a weld joint, the stress σ_h of the front fillet weld perpendicular to the weld length direction shall be calculated by Formula (8.4.3-3) and shall satisfy Formula (8.4.3-4):

$$\sigma_h = \frac{N}{h_e l_w} \quad (8.4.3\text{-}3)$$

$$\sigma_h \leq [\tau_h] \quad (8.4.3\text{-}4)$$

2) Under the action of tensile, compressive and shear forces through the centroid of a weld joint, the stress τ_h of the side fillet weld parallel to the weld length direction shall be calculated by Formula (8.4.3-5), and shall satisfy Formula (8.4.3- 6):

$$\tau_h = \frac{N}{h_e l_w} \quad (8.4.3\text{-}5)$$

$$[\tau_h] \leq [\tau_h] \quad (8.4.3\text{-}6)$$

3) Under the action of various forces, the stress at the position where both stresses σ_h and τ_h appear shall satisfy Formula (8.4.3-7):

$$\sqrt{\sigma_h^2 + \tau_h^2} \leq [\tau_h] \quad (8.4.3\text{-}7)$$

where

σ_h is the stress perpendicular to the length direction of the weld, which is calculated by the effective section of the weld (N/mm^2);

N is the force acting on the fillet weld (N);

h_e is the calculation height of fillet weld (mm), assuming that the fillet weld is a right-angle one (Figure 8.4.3), and the calculation height is $0.7h_f$, where h_f is the smaller leg size;

l_w is the calculation length of fillet weld (mm);

τ_h is the stress parallel to the length direction of the weld, which is calculated by the effective section of the weld (N/mm²);

$[\tau_h]$ is the allowable shear stress of fillet weld (N/mm²), which is taken as per Table 8.3.2-2 of this code.

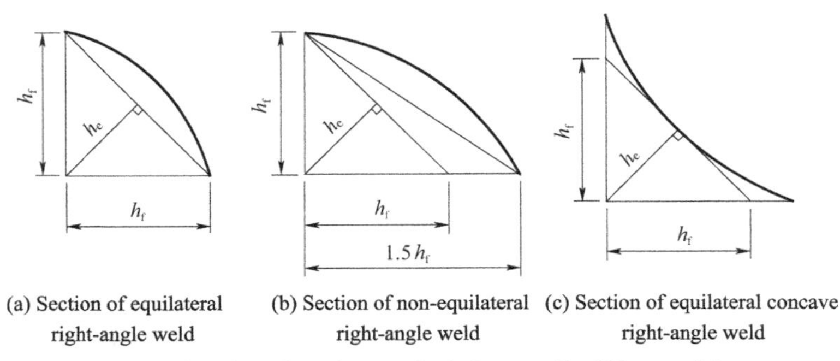

(a) Section of equilateral right-angle weld (b) Section of non-equilateral right-angle weld (c) Section of equilateral concave right-angle weld

Figure 8.4.3 Sections of right-angle fillet welds

3 The calculation length of welds shall meet the following requirements:

1) When the internal force is distributed along the full length of a side fillet weld, the calculation length of the weld is taken as its actual length minus $2h_f$.

2) The minimum calculation length of a fillet weld is $8h_f$.

3) For a side fillet weld, the maximum calculation length is $40h_f$ under dynamic load and $60h_f$ under static load. If the weld length is greater than that specified above, the excess shall not be considered in the calculation.

8.4.4 The strength calculation of bolt connections shall meet the following requirements:

1 The general purpose bolt connection shall meet the following requirements:

1) Grades A and B bolt connections may be used in structures under dynamic load.

2) Grade C bolt connections may only be used for tension connection or for temporary fixation during installation as the fit clearance between the bolt and the bolt hole is large.

3) For general purpose bolt connections, the tensile capacity and shear capacity of bolts shall be checked, and the compressive capacity of bolt holes shall be checked.

2 Friction-type high-strength bolt connections shall meet the following requirements:

1) For the connection under shear, the bearing capacity of a single friction-type high-strength bolt shall be calculated by the following formula:

$$[P]=\frac{Z_m \mu P_g}{n} \tag{8.4.4-1}$$

where

[P] is the allowable bearing capacity of a single standard-hole friction-type high-strength bolt (kN);

Z_m is the number of faying surfaces for force transmission;

μ is the mean slip coefficient of faying surface, which is selected as per Table 8.4.2 of this code;

P_g is the design pretension force of a high-strength bolt (kN), which is selected as per Table 8.4.4;

n is the safety factor, which is selected as per Table 8.3.1-1 of this code.

Table 8.4.4 Design pretension force P_g of a high-strength bolt (kN)

Performance level of bolt	Nominal diameter of bolt					
	M16	**M20**	**M22**	**M24**	**M27**	**M30**
8.8S	80	125	150	175	230	280
10.9S	100	155	190	225	290	355

2) The allowable bearing capacity of each high-strength bolt is still calculated by Formula (8.4.4-1) when the friction-type high-strength bolt connection bears both the shear force along the faying surface and the external tensile force along the bolt axis, but P_g in the formula shall be substituted by (P_g − 1.25P_t). P_t is the external tensile force, and shall not be greater than 0.7P_g.

3) The number Z of bolts in the friction-type high-strength bolt connection shall be calculated by the following formula:

$$Z=\frac{N}{[P]} \tag{8.4.4-2}$$

where

Z is the number of bolts in the friction-type high-strength bolt connection;

N is the internal force transmitted by the connection (kN);

$[P]$ is the allowable bearing capacity of a single standard-hole friction-type high-strength bolt (kN).

4) In the friction-type high-strength bolt connection, to ensure the specified pre-tension of high-strength bolt, the tightening torque for bolts and nuts and the operating method shall comply with the current sector standard JGJ 82, *Technical Specification for High Strength Bolt Connections of Steel Structures*.

8.4.5 When the foundation embedded parts of the hoist frame are embedded in phase Ⅰ or Ⅱ concrete, the compressive stress of the concrete shall be calculated based on the effective compression area of the embedded parts and shall be in accordance with Table 8.3.4 of this code.

8.5 Stability Calculation

8.5.1 The overall stability of flexural members shall meet the following requirements:

1 The overall stability of flexural members need not be checked in any of the following cases:

1) Walkways and slabs with a high stiffness are firmly connected to the compression flange of the flexural member, which can prevent the lateral displacement of the compression flange.

2) The ratio of the section height h of a box-section flexural member to the flange plate width b between the two webs is not greater than 3; or the member has a cross section whose size is sufficient to ensure the lateral stiffness, such as space truss.

3) For the flexural member with a constant cross-section rolled H-steel or welded I-steel simply supported at both ends and not twistable at the end, the ratio of the lateral support spacing l to the width b of the compression flange meets the following conditions: there is no lateral support and the load applied on the compression flange satisfies Formula (8.5.1-1); there is no lateral support and the load applied on the tension flange satisfies Formula (8.5.1-2); the midspan compression flange is provided with lateral support and satisfies Formula (8.5.1-3).

$$l/b \leq 13\sqrt{235/R_{eH}} \tag{8.5.1-1}$$

$$l/b \leq 20\sqrt{235/R_{eH}} \tag{8.5.1-2}$$

$$l/b \leq 16\sqrt{235/R_{eH}} \tag{8.5.1-3}$$

where

l is the lateral support spacing of compression flange (mm);

b is the width of compression flange (mm);

R_{eH} is the upper yield strength of material (N/mm^2).

2 The overall stability of flexural members that do not meet the above requirements shall be checked in accordance with the current national standard GB 50017, *Standard for Design of Steel Structures*.

8.5.2 The layout of web stiffeners of a composite beam (Figure 8.5.2) shall meet the following requirements:

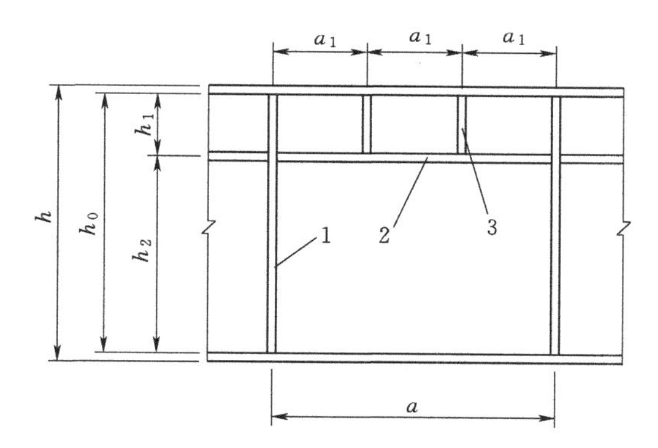

Key

1 transverse stiffener

2 longitudinal stiffener

3 short stiffener

Figure 8.5.2 Layout of web stiffeners of a composite beam

1 When the ratio of the height h_0 to the thickness t_w of the web satisfies Formula (8.5.2-1), transverse stiffeners shall be provided for beams under local compressive stress according to the detailing requirements. In the case of small local stresses, stiffeners need not be provided.

$$h_0/t_w \leq 80\sqrt{235/R_{eH}} \tag{8.5.2-1}$$

where

h_0 is the height of web (mm);

t_w is the thickness of web (mm);

R_{eH} is the upper yield strength of material (N/mm^2).

2 When the ratio of h_0 to t_w satisfies Formula (8.5.2-2), transverse stiffeners shall be provided.

$$h_0/t_w > 80\sqrt{235/R_{eH}} \quad (8.5.2\text{-}2)$$

3 When the compression flange can freely twist and the ratio of h_0 to t_w satisfies Formula (8.5.2-3), or the compression flange cannot freely twist and the ratio satisfies Formula (8.5.2-4), transverse stiffeners shall be provided, and longitudinal stiffeners shall be provided in the compression zone with high bending stress.

$$h_0/t_w > 150\sqrt{235/R_{eH}} \quad (8.5.2\text{-}3)$$

$$h_0/t_w > 170\sqrt{235/R_{eH}} \quad (8.5.2\text{-}4)$$

4 In no case shall the ratio of h_0 to t_w be greater than 250.

5 For the beam under high local compressive stress, short stiffeners should be provided in the compression zone if necessary.

6 Supporting stiffeners shall be provided at the beam support and at the upper flange under high point loads.

8.5.3 The local stability analysis of compression flange plate shall meet the following requirements:

1 When the ratio of free width b to the thickness δ of I-section flange plate in compression satisfies Formula (8.5.3-1), the local stability of the compression flange plate need not be checked.

$$b/\delta \leq 15\sqrt{235/R_{eH}} \quad (8.5.3\text{-}1)$$

where

b is the free width of I-section flange plate in compression;

δ is the thickness of I-section flange plate in compression;

R_{eH} is the upper yield strength of material (N/mm^2).

2 For a box-shaped beam, when the ratio of the distance b_0 between the centerlines of two webs to the thickness δ_y of the compression flange plate satisfies Formula (8.5.3-2), and the calculated compressive stress of the plate is not greater than 0.8[σ], the local stability of the compression flange plate need not be checked.

$$b_0 / \delta_y \leq 60\sqrt{235 / R_{eH}} \quad (8.5.3\text{-}2)$$

where

b_0 is the distance between the centers of two webs;

δ_y is the thickness of compression flange plate;

R_{eH} is the upper yield strength of material (N/mm^2).

8.5.4 The structural dimensions of stiffeners shall meet the following requirements:

1 Transverse stiffeners shall be set as follows:

1) When the web is only reinforced with transverse stiffeners, the spacing a of stiffeners shall satisfy Formulae (8.5.4-1) to (8.5.4-3):

$$a \leq \frac{615h_0}{\frac{h_0}{t_w}\sqrt{\eta\tau} - 765} \quad (8\ 5.4\text{-}1)$$

$$\tau = V/(h_w t_w) \quad (8\text{·}5.4\text{-}2)$$

$$\sigma = (My_1)/I \quad (8\ 5.4\text{-}3)$$

where

a is the spacing of stiffeners (mm), which is taken as $2h_0$ when the value calculated from the right side of Formula (8.5.4-1) is greater than $2h_0$ or the denominator is negative;

h_0 is the calculation height of web (mm);

t_w is the thickness of web (mm);

η is the amplification factor considering the influence of σ, which is selected in accordance with Table 8.5.4;

τ is the average shear stress of web generated by the maximum shear force in the beam section concerned (N/mm^2);

V is the shear force (N);

h_w is the height of web (mm);

σ is the bending compressive stress at the edge of the calculation height of the web at the same section of τ (N/mm^2);

y_1 is the distance from the compression edge of the calculation height of the web to the neutral axis (mm);

I is the gross inertia moment of beam section (mm^4).

Table 8.5.4 Amplification factor considering the influence of σ

$\sigma\left(\frac{h_0}{100t_w}\right)^2$	100	120	140	160	180	200
η	1.02	1.03	1.05	1.06	1.08	1.10
$\sigma\left(\frac{h_0}{100t_w}\right)^2$	220	240	260	280	300	320
η	1.13	1.16	1.19	1.24	1.29	1.35
$\sigma\left(\frac{h_0}{100t_w}\right)^2$	340	360	380	400	420	440
η	1.43	1.53	1.67	1.85	2.14	2.65

NOTE The factor η in the table is calculated by the formula: $\eta = \frac{1}{\sqrt{1-\left[\frac{\sigma}{475}\left(\frac{h_0}{100t_w}\right)^2\right]^2}}$

2) When the web of the beam is reinforced with both transverse and longitudinal stiffeners, the spacing a of transverse stiffeners shall still be determined by Formula (8.5.4-1), but h_0 shall be substituted by h_2, and η shall be 1.0.

3) Provided that the local stability of the web is met, the spacing a of transverse stiffeners of the web shall not be less than $0.5h_0$ but not be greater than $2h_0$.

4) The free width b_s of rectangular-section transverse stiffeners provided in pairs on both sides of the web shall satisfy Formula (8.5.4-4), and the thickness t_s of transverse stiffeners shall satisfy Formula (8.5.4-5):

$$b_s \geq \frac{h_0}{30} + 40 \qquad (8.5.4\text{-}4)$$

$$t_s \geq \frac{1}{15} b_s \sqrt{\frac{R_{eH}}{235}} \qquad (8.5.4\text{-}5)$$

where

b_s is the free width of transverse stiffener (mm);

h_0 is the height of web (mm);

t_s is the thickness of transverse stiffener (mm);

R_{eH} is the upper yield strength of steel (N/mm^2).

5) When rectangular-section transverse stiffeners are provided on one side of the web, the free width of stiffener shall be greater than 1.2 times b_s calculated by Formula (8.5.4-4) and the thickness shall not be less than 1/15 of b_s in order to obtain the same linear stiffness as that generated by transverse stiffeners on both sides.

6) When both transverse stiffeners and longitudinal stiffeners are provided for the web, the cross-sectional moment of inertia of transverse stiffeners shall satisfy the following formula in addition to the above provisions:

$$I_{s1} \geq 3h_0 t_w^2 \tag{8.5.4-6}$$

where

I_{s1} is the cross-sectional moment of inertia of transverse stiffener (mm^4);

h_0 is the height of web (mm);

t_w is the thickness of web (mm).

2 Longitudinal stiffeners shall be set as follows:

1) The distance h_1 from the longitudinal stiffener to the compression edge of the calculation height of the web shall be 1/5 to 1/4 of the height h_0 of the web.

2) For webs of flexural members, the cross-sectional moment of inertia of longitudinal stiffeners shall satisfy Formula (8.5.4-7) when the ratio of the spacing a of transverse stiffeners of the web to the height h_0 of the web is not greater than 0.85 and shall satisfy Formula (8.5.4-8) when the ratio is greater than 0.85:

$$I_{s2} \geq 1.5h_0 t_w^2 \tag{8.5.4-7}$$

$$I_{s2} \geq \left(2.5 - 0.45\frac{a}{h_0}\right)\left(\frac{a}{h_0}\right)^2 h_0 t_w^3 \tag{8.5.4-8}$$

where

I_{s2} is the cross-sectional moment of inertia of longitudinal stiffeners of the web (mm^4);

h_0 is the height of web (mm);

t_w is the thickness of web (mm);

a is the spacing of transverse stiffeners (mm).

3 For the stiffeners provided in pairs on both sides of the web, the cross-sectional moment of inertia shall be calculated by taking the web centerline as the axis. For the stiffeners provided on one side of the web, the cross-sectional moment of inertia shall be calculated by taking the edge of the web connected with the stiffeners as the axis.

4 The minimum spacing of short stiffeners shall be $0.75h_1$, the free width shall be taken as 0.7 to 1.0 times the free width of transverse stiffener, and the thickness shall not be less than 1/15 of the free width of short stiffener.

8.6 Stiffness Requirements

8.6.1 The stiffness of the frame beam system of the hoist whose drum and speed reducer are driven by the open gear pair shall meet the following requirements:

1 The maximum vertical static deflection of the main load-bearing beam system shall satisfy the following formula:

$$y_e \leq \frac{L}{2000} \tag{8.6.1-1}$$

where

y_e is the maximum vertical static deflection (mm);

L is the support span (mm).

2 The maximum vertical static deflection of other beam systems shall satisfy the following formula:

$$y_e \leq \frac{L}{1000} \tag{8.6.1-2}$$

8.6.2 The stiffness of the frame beam system of the hoist whose drum and speed reducer are driven by closed gears shall meet the following requirements:

1 When the low-speed end of the speed reducer is connected to the drum using the ball-and-socket hinge style drum coupling, spherical roller coupling or drum gear coupling, the maximum vertical static deflection of the frame structure shall satisfy the following formula:

$$y_e \leq \frac{L}{1000} \tag{8.6.2-1}$$

2 When the low-speed end of the speed reducer is connected to the drum by other methods, the maximum vertical static deflection of the main load-bearing beam system of the frame structure shall satisfy the

following formula:

$$y_e \leq \frac{L}{2000} \tag{8.6.2-2}$$

8.6.3 The stiffness of the frame beam system of the hoist with a disc safety brake set on the drum shall meet the following requirements:

1 The maximum vertical static deflection of the main load-bearing beam system shall satisfy the following formula:

$$y_e \leq \frac{L}{2000} \tag{8.6.3-1}$$

2 The maximum vertical static deflection of other beam systems shall satisfy the following formula:

$$y_e \leq \frac{L}{2000} \tag{8.6.3-2}$$

8.7 Detailing Requirements

8.7.1 The structural details shall meet the following requirements:

1 The main load-bearing structures shall be simple in construction, and clear in stress conditions, and the impact of stress concentration shall be minimized.

2 The structures shall be designed to facilitate manufacturing, inspection, transportation, installation, and maintenance. The structures exposed to open air shall be protected from water accumulation.

3 The thickness of steel plates and section steel legs of main load-bearing structures shall not be less than 5 mm.

4 Different connection modes are allowed for the main load-bearing members at different connections, but not allowed at the same connection.

5 The butt welds of the web and flange plate of the welded beam should not be on the same section, and the spacing shall not be less than 200 mm. The transverse stiffeners shall be staggered with the web butt welds that are parallel to the transverse stiffeners, and the spacing shall not be less than 200 mm.

8.7.2 Welded connections shall meet the following requirements:

1 The deposited metal shall be compatible with the base metal. When the steels with different strengths are welded, the welding materials compatible with low-strength steels may be used.

2 Butt welds shall meet the following requirements:

1) The groove type of butt welds shall comply with the current national standards GB/T 985.1, *Recommended Joint Preparation for Gas Welding, Manual Metal Arc Welding, Gas-Shield Arc Welding and Beam Welding*; and GB/T 985.2, *Recommended Joint Preparation for Submerged Arc Welding*.

2) For the connection of plates with unequal thickness or width in the main load-bearing structure, the transition of butt welding (Figure 8.7.2-1) shall adopt a slope not steeper than 1 : 4 from one side or both sides.

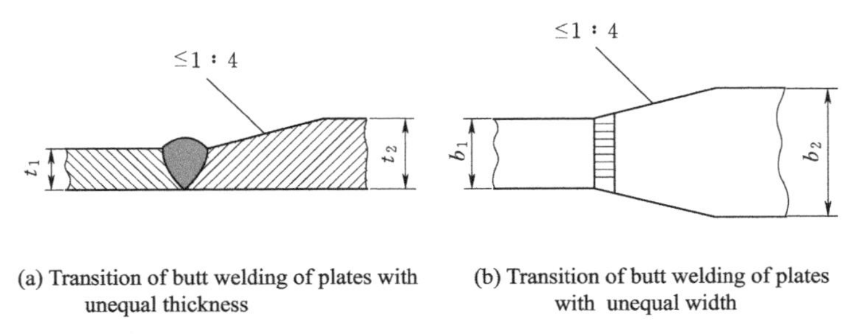

(a) Transition of butt welding of plates with unequal thickness

(b) Transition of butt welding of plates with unequal width

Figure 8.7.2-1 Transition of butt welding

3 Fillet welds shall meet the following requirements:

1) The minimum height h_f of fillet weld legs shall be in accordance with Table 8.7.2.

Table 8.7.2 Minimum height h_f of fillet weld leg

Thickness of thicker weldment *t* (mm)	h_f (mm)	
	Ordinary carbon steel welding	**Low alloy steel welding**
$t \leq 10$	4	6
$10 < t \leq 20$	6	8
$20 < t \leq 30$	8	10

2) For general fillet welds, the size of the weld leg shall not be greater than 1.2 times the thickness of the thinner weldment.

3) For fillet welds at the edge of the member, the leg of the fillet weld (Figure 8.7.2-2) shall also meet the following requirements: when t_1 is not greater than 6 mm, h_f is not greater than t_1; when t_1 is greater than 6 mm, h_f is t_1 – (1 to 2) mm.

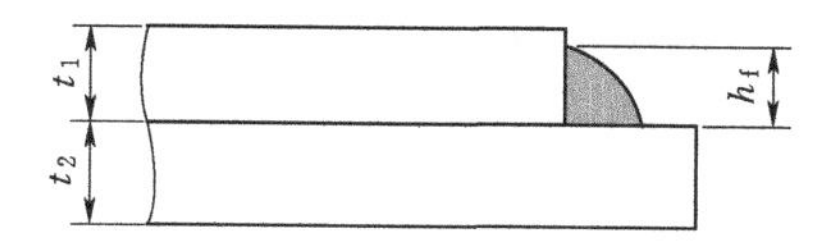

Figure 8.7.2-2 Leg of fillet weld

4 The weld shape shall meet the following requirements:

1) For the main bearing structure subject to dynamic loads, the surface of the fillet weld shall be slightly concave or straight, provided that the calculated cross-sectional area of the weld under shear is guaranteed.

2) The ratio of right-angle sides is 1 : 1 for side welds, and 1 : 1.5 for end welds. The longer side of welds is along the load direction.

8.7.3 Bolted connections shall meet the following requirements:

1 For large structures that need to be integrally spliced at site, high-strength bolts should be used for connection.

2 The number of bolts for each member shall not be less than 2 at the node or on one side of the joint, and the number of bolts in each row should not be more than 5 along the stress direction.

3 Grade C bolts are only allowed to be used for the connection of secondary members or for temporary connections.

4 Effective measures against nut loosening shall be taken for general purpose bolt connections.

5 When reamed hole bolts are used, the fit between the hole and the bolt shall not be inferior to H11/h9.

6 When high-strength bolts are used for connection, high-strength washers shall be provided at the heads of nuts and bolts.

7 Standard holes shall be used for high-strength bolt connections, and the hole diameter for high-strength bolt connection shall be in accordance with Table 8.7.3-1.

Table 8.7.3-1 Hole diameter for high-strength bolt connection (mm)

Nominal diameter of bolt	M16	M20	M22	M24	M27	M30
Hole diameter	17.5	22	24	26	30	33

8 The holes for high-strength bolt connections shall comply with the

current sector standard JGJ 82, *Technical Specification for High Strength Bolt Connections of Steel Structures*.

9 The limit size for bolt arrangement shall be in accordance with Table 8.7.3-2.

Table 8.7.3-2 Limit size for bolt arrangement

<table>
<tr><th>Item</th><th colspan="3">Location and direction</th><th>Maximum allowable distance</th><th>Minimum allowable distance</th></tr>
<tr><td rowspan="5">Center-to-center distance</td><td colspan="3">Outer row, perpendicular to or along the direction of internal force</td><td>$8d_0$ or $12t$</td><td rowspan="5">$3d_0$</td></tr>
<tr><td rowspan="3">Middle row</td><td colspan="2">Perpendicular to the direction of internal force</td><td>$16d_0$ or $24t$</td></tr>
<tr><td rowspan="2">Along the direction of internal force</td><td>Compression member</td><td>$12d_0$ or $18t$</td></tr>
<tr><td>Tension member</td><td>$16d_0$ or $24t$</td></tr>
<tr><td colspan="3">Along the diagonal direction</td><td>–</td></tr>
<tr><td rowspan="4">Distance from center to member edge</td><td colspan="3">Along the direction of internal force</td><td rowspan="4">$4d_0$ or $8t$</td><td>$2d_0$</td></tr>
<tr><td rowspan="3">Perpendicular to the direction of internal force</td><td colspan="2">Sheared edge or manual gas-cut edge</td><td>$1.5d_0$</td></tr>
<tr><td rowspan="2">Rolled edge, automatic gas-cut or sawn edge</td><td>High-strength bolt</td><td rowspan="2">$1.2d_0$</td></tr>
<tr><td>Other bolts</td></tr>
</table>

NOTES

1 d_0 is the hole diameter of the bolt, and t is the thickness of the thinner plate at the outer layer.

2 The maximum distance between bolts connecting the edge of the steel plate and the rigid member may be selected based on the values in the middle row.

9 Electrical Design

9.1 General Requirements

The electrical equipment shall comply with the current national standard GB/T 5226.32, *Electrical Safety of Machinery—Electrical Equipment of Machines—Part 32: Requirements for Hoisting Machines*.

9.2 Power Source and Power Supply

9.2.1 The power source shall be low-voltage three-phase 3Φ + PE, 50Hz/60Hz AC power, which is not higher than 1000 V.

9.2.2 Under normal operating conditions, the voltage fluctuation at the feeder access point of the hoist shall not exceed ±10 % of the rated voltage.

9.2.3 The hoist shall be provided with protective earthing conductor, which shall be reliably earthed.

9.3 Power Distribution System

9.3.1 The hoist shall be provided with a main disconnector that can cut off all power supplies.

9.3.2 The main power circuit shall be provided with a main circuit breaker. The main circuit breaker shall be designed with the function of electromagnetic tripping, its rated current shall be greater than the rated operating current of the hoist, and the current setting of the electromagnetic tripping shall be greater than the maximum operating current of the hoist. The main circuit breaker shall have the arc breaking capacity to disconnect the short-circuit current that occurs to the hoist.

9.3.3 The power circuit should be provided with a contactor capable of disconnecting the power line.

9.3.4 An emergency stop button shall be provided for the hoist to disconnect the main power supply and the brake power supply in case of emergency. The emergency stop button shall be red and be of the non-self-resetting type.

9.3.5 The branch lines of the service, control and auxiliary power supplies shall be designed with overcurrent protection, which may be realized through circuit breakers, fuses or overcurrent protection relays. The three-phase power circuit, if fuse protected, shall have the function of open-phase protection.

9.4 Electrical Protection

9.4.1 The motor shall have one or several of the following protections depending on the motor type and its control mode:

1 Overcurrent protection of instantaneous or inverse time actions. The current setting value for instantaneous action is about 1.25 times the maximum starting current of the motor.

2 Thermal sensing elements in the motor.

3 Thermal overload protection.

9.4.2 The external power supply line to the hoist shall be provided with the overcurrent protection against short circuiting or line-to-earth fault. In the case of short circuiting or line-to-earth fault, the instantaneous protection device shall be able to disconnect the line. For control or auxiliary lines with a small cross-sectional area and long external extension, if the estimated earthing current cannot reach the instantaneous tripping current, the thermal tripping function shall be added to ensure that conductor insulation is not burnt out due to line-to-earth fault.

9.4.3 The hoist shall be provided with the phase-fault or open-phase protection.

9.4.4 The drive mechanism of the hoist shall be provided with zero position protection. The mechanism shall not act automatically when the power supply is restored after the operation stops due to fault or voltage loss, and the mechanism shall not restart until the controller is manually zeroed.

9.4.5 The hoist shall be provided with voltage loss protection. When the power supply for the hoist is interrupted, the electric equipment that might cause safety problem or should not re-start automatically shall all be in a power-off state to prevent the electric equipment from starting automatically after the power supply is restored.

9.4.6 For an important lifting mechanism whose overspeed under load is likely to cause danger, an overspeed switch shall be provided. The setting value of the overspeed switch should be 1.25 to 1.4 times the rated lowering speed. For the lifting mechanism with a safety brake, an overspeed detector shall be provided on the low-speed shaft, and the overspeed setting value should not be greater than 1.5 times the specified lowering speed.

9.4.7 The hoist shall be provided with accidental power loss protection for the motor stator. When the motor is out of control due to the failure of speed regulating device or forward and reversing contactors, the brake shall be applied immediately.

9.4.8 The earthing protection shall meet the following requirements:

1 The main metal structure of the hoist shall be reliably connected to the

earthing network.

2 The metal enclosures, conduits, supports and metal trunking of all electrical equipment of the hoist shall be reliably earthed. Special earthing conductors should be used to ensure reliable earthing of electrical equipment.

3 The conductance of earthing conductors and earthing facilities should not be less than 1/2 times the maximum phase conductance in the line.

4 The earthing conductors shall not be used as current-carrying neutral lines.

9.4.9 For hoists installed outdoors and higher than the surrounding ground surface, consideration shall be given to preventing lightning strike from causing damage to high-position parts and injury to personnel.

9.4.10 The live parts of the hoist electrical control equipment , if exposed and accessible, shall be protected against electric shock.

9.4.11 The hoist may be provided with a separate lighting transformer. The lighting transformer shall be an isolating transformer rather than an autotransformer. One end of the secondary side of the lighting transformer shall be earthed. The circuit breaker and leakage protection switch shall be provided for the main power switch of the lighting line.

9.4.12 The safety lighting voltage shall not be greater than 50 V.

9.4.13 The hoist shall be provided with a load limiter and a height indicator, which shall comply with Articles 7.3.9 and 7.3.10 of this code.

9.5 Electrical Control

9.5.1 Control elements shall meet the following requirements:

1 The contactors shall comply with the current national standard GB/T 14048.4, *Low-voltage Switchgear and Controlgear—Part 4-1: Contactors and Motor-Starters—Electromechanical Contactors and Motor-Starters (Including Motor Protector)*. Electrical and mechanical interlocks shall be provided between the reversing contactor and the other contactors that might cause short circuit when they are closed simultaneously.

2 The design and type selection of programmable logic controller may be conducted according to the number of inputs and outputs and voltage class of the switching values, number of inputs and outputs of analog values, and other special functional requirements, and the redundant

system may be used if particularly high reliability is required. The direct relay protection interlock circuit shall be provided for the interlock signals used for safety protection.

3 Resistors shall meet the following requirements:

1) For motors with different cyclic duration factors, the general-purpose resistors with different parameters should be selected; for motors with different but similar cyclic duration factors, resistors of the same specifications may be selected.

2) For the starting resistors at all levels, the allowable resistance deviation between the selected value and the calculated value shall be ± 5 %. The allowable resistance deviation between the selected value and the calculated value of the resistor at a particular level shall be ± 10 %. However, the allowable deviation between the selected value and the calculated value of the total resistance of each phase shall be ± 8 %.

3) The heating capacity of a resistor is generally selected on the basis of repeated short-time duty. The time of one cycle is set as 60 s, and the cyclic duration factors are 100 %, 70 %, 50 %, 35 %, 25 %, 17.5 %, 12.5 %, 8.8 %, 6.25 % and 4.4 %, respectively. The heating capacity of the common cascaded resistor shall be selected on the basis of long-time duty.

4) Different cyclic duration factors may be selected for different connection conditions for the resistors at each level. The same resistor element may have different allowable current values for different cyclic duration factors, but the allowable current value of the selected resistor element shall not be less than the rated current of the motor rotor.

5) Frequency sensitive resistors shall not be selected for the lifting mechanism.

6) When the brake unit is used in the variable frequency speed regulating system, the cyclic duration factor of the resistor of the lifting mechanism shall be 100 %, and the power value of the resistor shall not be less than the rated feedback power at the time of lowering.

7) The resistors shall be provided with a protective cover. The protection degree shall not be lower than IP10 as specified in the current national standard GB/ T 4208, *Degrees of Protection*

Provided by Enclosure (IP Code) when it is used indoors, and shall not be lower than IP13 when it is used outdoors.

8) The resistors shall be securely installed. Generally, up to 4 resistor boxes may be stacked directly. When there are more than 4 boxes, the number of stacked boxes may be increased, provided that stable heat dissipation and temperature rise requirements are met.

9.5.2 Control panels/cabinets shall meet the following requirements:

1 For locally controlled hoists, all actions of mechanisms shall be controlled through the operating elements such as buttons and switches on the control panel.

2 Except for direct control of the motor, the voltage in the control device shall not be greater than 250 V.

3 The enclosure of the control device should be made of completely insulated materials or materials with insulating protective layers, and the metal enclosures or accessible metal parts shall be earthed separately.

4 The control panel/cabinet of the outdoor hoist shall be provided with IP protection. When installed in an unsheltered place, the protection degree of the enclosure shall not be lower than IP54 as specified in the current national standard GB/T 4208, *Degrees of Protection Provided by Enclosure (IP Code)*; when installed in a sheltered place, the protection degree of the enclosure shall not be lower than IP43.

5 The metal enclosure of the equipment need to be provided with protective earthing screws or nuts welded on the enclosure and marked with protective earthing symbols at conspicuous places. If there are electrical elements on the door, dedicated earthing conductors shall be provided, and the door shall be lockable.

6 The control panel/cabinet shall be securely installed, with a clearance of at least 600 mm in front of the enclosure or cabinet, and there shall be no obstacles on the ground.

9.5.3 Control system shall meet the following requirements:

1 The performance of the control system shall meet the following requirements:

1) The hoist shall be able to operate reliably at 110 % of the rated load.

2) When rated load is lifted while the voltage fluctuation of the power

supply system is −10 % of the rated value, the control system shall ensure the normal operation of the lifting mechanism without hook slipping regardless of the position of the load.

2 When the control power is supplied by the transformer, the voltage on the secondary side shall not be greater than 250 V to prevent overload or short circuit of non-common pole lines.

9.5.4 Electrical drive scheme shall meet the following requirements:

1 The AC drive control system may be used for the hoist, and shall meet the following requirements:

1) The control scheme such as contactor, variable frequency, multi-speed motor and double motor should be used.

2) The capacity of main components such as motor and resistor of the speed regulating system shall be selected and verified according to the actual operating conditions.

3) Variable frequency speed regulation may achieve constant torque speed regulation at frequencies below the rated frequency and constant power speed regulation at frequencies above the rated frequency. The maximum frequency of low-intensity magnetic speed rising of constant power speed regulation should not be greater than 2 times the rated frequency.

4) When a variable frequency speed regulation is used for the lifting mechanism, a closed-loop control shall be used if the constant torque speed regulation range is greater than 1 : 10.

5) The maximum output current of the frequency converter shall not be less than the required maximum starting current, and the rated output current shall not be less than the operating current of the motor at the rated load. The design frequency of the frequency converter shall be so determined as to avoid resonance with the hoist structure.

6) The variable frequency speed regulation drive system should be used for the hoist with speed regulation or electrical synchronization operation requirements.

2 For the lifting mechanism driven by the wound-rotor asynchronous motor controlled by the control panel, electrical braking shall be available for the lowering and decelerating process.

3 When the frequency converter speed regulating drive system is used, regenerative braking or energy consumption braking may be used.

9.5.5 The control of brakes shall meet the following requirements:

1 Supporting brakes shall meet the following requirements:

1) Measures shall be taken for the control of the supporting brake to prevent uncontrolled movement during starting and braking.

2) In the case of electrical braking, the mechanical braking shall be applied after the electrical braking.

3) The brake shall not be applied when the motor is energized, except for transient state.

2 For the lifting mechanism with a safety brake, after the supporting brake acts, the safety brake action after delay, and the delay time is adjustable. The safety brake shall act promptly in the case of emergency braking.

3 The brake shall be applied in a timely manner in case of system power outage to prevent damage to the mechanical equipment.

9.5.6 The synchronous control of double lifting points shall meet the following requirements:

1 When a mechanical synchronous shaft is provided, the following requirements shall be met:

1) For the AC drive system with the contactor starting control mode, the control may be realized by simultaneously starting or stopping the motors.

2) For the AC drive system controlled by a frequency converter, the control may be realized by simultaneously starting or stopping the motors, and each motor shall be provided with a separate overcurrent protective device.

3) When a set of AC drive system controlled by the frequency converter is used for each lifting point, the torque master-slave control mode or the control mode with double lifting point torque balance shall be used.

4) When two wound-rotor motors are driven coaxially, the rotor winding shall be provided with normally connected resistors to facilitate load balance.

2 During electrical synchronization, a set of variable frequency speed regulating device or other speed regulating devices with the same functions shall be used for each lifting point. The synchronous control of two lifting points shall be realized by detecting the travel difference between two lifting points and adjusting the speed of either of the motors. When the travel difference between two lifting points is less than 10 mm, no adjustment is made, and when it is greater than 20 mm, the motors are switched off to stop operation.

9.6 Working Environment of Electrical Equipment

9.6.1 The electrical equipment shall be free of oil drops resulting from installation or operation of any lubricating system, hydraulic system or other oil-containing devices; otherwise, the electrical equipment shall be protected.

9.6.2 The electrical equipment of the hoist shall be able to work normally at the operating ambient temperature and humidity. Cooling or heating measures shall be taken when the ambient temperature and humidity cannot meet the working requirements of the selected electrical equipment.

9.6.3 When the ambient air temperature reaches +40 °C, the relative humidity shall not be greater than 50 %; when the temperature is relatively low, the relative humidity may be increased, but the occasional condensation caused by temperature change shall be taken into account. When the ambient air temperature is not greater than +25 °C, the relative humidity is allowed to be up to 100 % for a short duration.

9.6.4 If the hoist is used at an altitude above 1,000 m, the specific data of the electrical equipment shall be as specified by the product standards.

9.6.5 If a hoist room is arranged, the hoist room shall be provided with auxiliary lighting that is not controlled by the main power supply, with an average illumination not less than 30 Lx.

9.7 Wires, Cables and Their Laying

9.7.1 Wires and cables should be copper-cored multi-strand flexible, and shall be selected according to the voltage class, ambient temperature and laying method. Control and signal cables shall be shielded cables.

9.7.2 The cross-sectional areas of wires and cables shall be selected according to the following requirements:

1 The cross-sectional area of the conductor shall be determined according to the load current borne by the conductor, the allowable voltage drop of the line, the operating ambient temperature and the mechanical strength required by the laying method.

2 The cross-sectional areas of external wires and cables of the hoist shall be selected as follows: for multi-strand single-core conductors, the cross-sectional area of a single core wire shall not be less than 1.5 mm^2; for multi-strand multi-core cables, the cross-sectional area of a single core wire shall not be less than 1 mm^2. The cross-sectional areas of connecting wires of electronic devices, hydraulic servo mechanisms, and detection and sensing elements are not particularly specified.

9.7.3 The load current of conductors shall meet the following requirements:

1 The load current of the conductor is the operating current of the electrical device.

2 The rated current-carrying capacity of the conductor shall not be less than the rated operating current of the connected electrical device.

3 The load current of the conductor that supplies power to the motor is the rated operating current of the motor.

4 The load current of the main power conductor of the hoist shall be calculated by the following formula:

$$I_W = I_N + I_{AUXI} \tag{9.7.3}$$

where

I_W is the load current of the main power conductor (A);

I_N is the rated operating current of the motor of the lifting mechanism (A);

I_{AUXI} is the operating current of main circuit and control circuit of auxiliary electrical equipment (A).

5 The cyclic duration factor of the load current I_W of the main power conductor is the same as that of the rated operating current of the motor, and the cyclic duration factors shall be unified in calculation based on the principle of equivalent heating.

9.7.4 The voltage drop shall meet the following requirements:

1 In the AC power supply system, the total voltage drop of the power supply from the low-voltage busbar of the power supply transformer to the terminal of any motor of the hoist shall not exceed 15 % of the rated voltage at the peak current.

2 The internal voltage drop of the hoist shall meet the following requirements:

1) The internal voltage drop of the hoist shall not exceed 5 %.

2) Under special circumstances, the fluctuation range of the power supply voltage and the internal voltage drop of the hoist may be agreed upon by the manufacturer and the user.

3 When the inductance per unit length of the conductor is ignored, the voltage drop of the AC conductor of the hoist is calculated by the following formulae:

$$\Delta U = \frac{\sqrt{3} L I_M \cos\varphi}{S\gamma} \tag{9.7.4-1}$$

$$I_M = I_{max} + I_{AUXI} \tag{9.7.4-2}$$

where

ΔU is the voltage drop of conductor (V);

L is the effective length of conductor (m);

I_M is the maximum total operating current of hoist (A);

$\cos\varphi$ is the power factor;

S is the cross-sectional area of conductor (mm^2);

γ is the conductivity of conductor [m/(Ω · mm^2)], taken as 54.34 m/(Ω · mm^2) for copper;

I_{max} is the maximum operating current of hoist (A);

I_{AUXI} is the operating current of main circuit and control circuit of auxiliary electrical equipment (A).

4 The maximum operating current of the hoist shall meet the following requirements:

1) The maximum operating current of a single motor is the maximum starting current selected in design, which may be 2.2 to 2.5 times the rated operating current I_N of the motor for the wound-rotor motor and 1.5 to 1.8 times I_N for the squirrel cage motor controlled by a frequency converter, and shall be obtained by consulting the catalog of the motor for the line-start squirrel cage motor.

2) The maximum operating current of a single hoist is calculated by the following formula when the motors of the lifting mechanism are all in the starting state:

$$I_{max} = K I_N \tag{9.7.4-3}$$

where

I_{max} is the maximum operating current of hoist (A);

K is the multiple of starting current of lifting mechanism motor;

I_N is the rated operating current of motor at rated load (A).

5 When calculating the voltage drop, the power factor cosφ at the time the motor is started, should be selected as follows:

1) 0.65 for wound-rotor motors.

2) 0.8 to 0.82 for squirrel cage motors with a variable frequency speed regulation.

3) 0.5 for line-start squirrel cage motors.

9.7.5 The current-carrying capacity of conductors should be corrected in accordance with Appendix G of this code.

9.7.6 Conductors shall be laid according to the following requirements:

1 The conductors shall be laid in metal trunking and metal conduits where physical damage might occur. Effective corrosion control measures shall be taken at the cable connections where chemical corrosion might occur. The cables shall be protected from oil pollution if any. The cables shall be prevented from being worn at the outlets of the trunking and conduit.

2 When cables are laid and fixed, the bending radius shall not be less than 5 times the outer diameter of the cable and shall not be less than the data provided by the manufacturer.

3 Three-phase or single-phase AC power lines shall be arranged in the same conduit.

4 Junction boxes shall be provided at the connection and branching points of conductors, and the protection degree shall be suitable for the surrounding environment.

5 For wiring terminals of electrical equipment such as control panels and junction boxes, if unexpected connection between the wiring terminals might cause equipment damage, the related terminals shall be enough spaced.

6 The conductors with different power supply voltage classes may be laid in the same trunking or conduit if the insulation withstand voltages of conductors are all greater than the highest power supply voltage class. The safety lighting power supply cable shall be laid separately.

7 The power cables for the output of the frequency converter should be laid separately from the control cables. Parallel routing shall be avoided when conditions permit. The cables shall be away from the equipment with electronic devices or sensing and detecting elements. Shielding measures shall be taken for important signals.

8 Except for conductors directly connected to crimp terminals, copper cold-pressed terminals shall be used at both ends of the conductors.

10 Safety

10.1 Markings, Nameplates and Safety Signs

10.1.1 The hoist shall be provided with markings, nameplates, and safety signs.

10.1.2 The rated lifting force and holding force of the hoist shall be permanently marked in a plainly visible place.

10.1.3 Each hoist shall be provided with a nameplate at an appropriate location, indicating at least the following:

1 Name of manufacturer.

2 Product name and model.

3 Main performance parameters.

4 Ex-factory serial number.

5 Date of manufacture.

10.1.4 Obvious textual warning safety signs shall be provided at appropriate positions of the hoist. Safety signs shall be provided at the hazardous positions of the hoist. The safety signs shall comply with the current national standards GB 2894, *Safety Signs and Guideline for the Use*; and GB/T 15052, *Cranes—Safety Signs and Hazard Pictorials—General Principles*. The colors of safety signs shall comply with the current national standard GB 2893, *Safety Colours*.

10.2 Structural Safety Requirements

10.2.1 The hoist operation cab shall meet the following requirements:

1 The operation cab shall have sufficient space to meet the ergonomic requirement. The operator should be seated or may stand when necessary.

2 The operation cab shall have safety exits. When a door is provided for the operation cab, the door shall be prevented from opening automatically during the operation of hoist. The sliding door and outward-opening door of the operation cab shall lead to the horizontal platform at the same elevation. When there is no platform outside the operation cab, the door should be opened inward.

3 The illuminance on the working surface of the operation cab shall not be less than 30 Lx.

4 Important operation indicators shall provide conspicuous displays

and installed in positions convenient for operators to observe. The indicators, alarm lights and emergency stop button shall have durable and legible markings. The indicators shall have appropriate ranges and shall be easy to read. The alarm lights shall have appropriate colors, and hazard warning lights shall be in red.

10.2.2 Passages and platforms shall meet the following requirements:

1 The access to operate and maintain the hoist shall be provided, including stairs, platforms, walkways or ladders, and guardrails shall be set as appropriate.

2 The clear height of inclined ladders, passages and platforms shall not be less than 1.8 m. The width of passages and platforms near the moving parts shall not be less than 0.5 m.

10.2.3 Inclined ladders and vertical ladders shall meet the following requirements:

1 Inclined ladders or vertical ladders shall be provided for passageways with a height difference of over 0.5 m, steps may be provided on the vertical plane with a height not exceeding 2.0 m, and handrails shall be provided on both sides of the steps.

2 The inclination of inclined ladders should not exceed 65°, the clear width of the steps shall not be less than 0.32 m, the step riser height should be 0.18 m to 0.25 m, the tread depth of the step shall not be less than its riser height, and the riser height and tread depth shall be the same for the steps consecutively arranged. Guardrails shall be provided on both sides of the inclined ladders, with a spacing no less than 0.6 m, and a height no less than 1.0 m.

3 The stringers on both sides of a vertical ladder shall be no less than 0.4 m apart and shall be 1.0 m higher than the top rung. The width of rungs between two stringers shall not be less than 0.3 m. The rungs shall be equally spaced, preferably at 0.23 m to 0.30 m. The rungs shall not be less than 0.15 m away from the fixed structural members. The rungs shall be capable of supporting at their center a vertical force of 1200 N distributed over 0.1 m without permanent deformation.

4 Vertical ladders with a height above 2.0 m shall be provided with hoop guards, and the hoop guards shall be installed from a height of 2.0 m and should have a diameter of 0.6 m to 0.8 m. The hoop guards shall be connected by longitudinal bars. In all cases, a bar shall be fixed in a position diametrically opposite to the vertical centerline of the ladder.

The hoop guard shall allow application at any point of a vertical force of 1,000 N distributed over 0.1m without permanent deformation.

10.3 Mechanical Safety Requirements

10.3.1 Wire ropes shall meet the following requirements:

1 The diameter of the wire rope should not be less than 6 mm.

2 When the load for a single lifting point is borne by multiple wire ropes, devices shall be provided to effectively ensure the load balance of each wire rope, and the coefficient of uniformity shall be considered.

3 The wire ropes shall not be spliced.

10.3.2 Drums shall meet the following requirements:

1 Wire ropes shall be arranged neatly on the drum in order.

2 The drums for multi-layer winding should be provided with flange guide plates to facilitate the automatic layered-winding of the wire rope.

3 The multi-layer winding drum shall be provided with flanges to prevent the wire rope from slipping at the end of the drum. The flange shall extend beyond the outer surface of the outermost wire rope by no less than 1.5 times the diameter of the wire rope.

4 When the movable pulley block is at the lower limit position, the fixed wraps and at least 2 safety wraps of the wire rope shall be wound on the drum, and the excess wire rope shall be cut off. When the movable pulley block is at the upper limit position, the rope winding allowance of not less than one full wrap shall be left on the drum.

5 The fixing device at the tail end of the wire rope on the drum shall be safe and reliable and have the functions of anti-loosening or self-tightening.

6 When the structure without the through supporting shaft is used in the drum body, the drum body shall be rolled with steel plates. The butt welds of each section of the welded drum shall be subjected to nondestructive testing as required.

10.3.3 Pulleys shall meet the following requirements:

1 The movable pulley block shall be provided with devices or structures to prevent the wire rope from falling out of the rope groove. The clearance between the side plate and arc top plate of the pulley cover

and the pulley body shall comply with the current national standard GB/T 6067.1, *Safety Rules for Lifting Appliances—Part 1: General*, and shall not be greater than 0.5 times the diameter of the wire rope.

2 Devices or structures shall be provided to prevent the wire rope from falling out of the rope groove when the wire rope on the non-movable pulley or equalizer pulley might fall out of the rope groove and cause adverse consequences.

3 The movable pulley block immersed in water shall be provided with an enclosed pulley cover.

10.3.4 Brakes shall meet the following requirements:

1 The lifting mechanism shall be provided with a normally closed mechanical brake.

2 The brake shall automatically support the gate when the gear shifting mechanism is shifted to the middle position.

3 The brake shall be adjustable and easy to check, and the brake lining shall be easy to replace.

4 The brake that can compensate for the wear of the brake lining shall be selected.

10.3.5 Effective measures shall be taken for fasteners such as bolts and screws to prevent loosening.

10.4 Electrical Safety Requirements

10.4.1 The hoist shall have signals indicating the on and off status of the master power supply, and fault or alarm signals shall be set if necessary. The signal indicators shall be set at the places audible and visible to the operator.

10.4.2 For hoists installed higher than the surrounding ground surface, consideration shall be given to preventing lightning strike from causing damage to high-position parts and injury to personnel.

10.5 Safety and Arrangement of Control and Manipulation

10.5.1 The control and manipulation system shall be designed and arranged to avoid the possibility of misoperation and to ensure the safe and reliable operation of the hoist in regular service.

10.5.2 The emergency stop button of the master power supply of the hoist shall comply with Article 9.3.4 of this code.

10.5.3 Text marks or symbols shall be imprinted on or near each control device to distinguish the functions and clearly indicate the movement direction of the manipulated hoist.

10.6 Requirements for Setting Safety Protection Devices

10.6.1 The requirements for load limiters shall be in accordance with Article 7.3.9 of this code.

10.6.2 The upper and lower position limiters shall comply with Item 2 of Article 7.3.10 of this code.

10.6.3 Safety protection devices shall be provided during normal operation or maintenance. Protective covers or guardrails shall be provided for exposed moving parts of the hoist that might hurt people. Rainproof measures shall be taken for the electrical equipment on the hoist operating in the open air.

10.6.4 Effective measures shall be taken in the design to prevent the parts and components of the hoist from falling.

10.7 Fire Protection

Fire protection facilities shall be provided according to the project rank. The configuration of fire protection facilities shall comply with the current national standard GB 50872, *Code for Fire Protection Design of Hydropower Projects.*

10.8 Operation and Maintenance Manuals

10.8.1 The hoist supplier shall provide the user with a hoist driving manual to guide the safe use of the hoist, and its content shall comply with the current national standard GB/T 17909.1, *Cranes—Crane Driving Manual—Part 1: General.*

10.8.2 The hoist supplier shall provide the user with a hoist maintenance manual to guide the normal maintenance of the hoist, and its content shall comply with the current national standard GB/T 18453, *Cranes—Maintenance Manual—Part 1: General.*

Appendix A Parameters of Hoisting Force, Lift, and Hoisting Speed Series

A.0.1 The parameters of hoisting force series should be in accordance with Table A.0.1.

Table A.0.1 Parameters of hoisting force series (kN)

50	63	80	100	125	160	200	250	320	400
500	630	800	1000	1250	1600	1800	2000	2200	2500
2800	3200	3600	4000	4500	5000	5600	6300	7000	8000
9000	10000	11000	12500	14000	16000	18000	20000	–	–

A.0.2 The parameters of lift series should be in accordance with Table A.0.2.

Table A.0.2 Parameters of lift series (m)

6	7	8	9	10	11	12	13	14	15	16	18	20
22	24	26	28	30	32	34	36	38	40	45	50	55
60	65	70	75	80	90	100	110	120	130	140	150	160

A.0.3 The parameters of hoisting speed series should be in accordance with Table A.0.3.

Table A.0.3 Parameters of hoisting speed series (m/min)

0.2	0.3	0.5	0.8	1.0	1.2	1.6	2.0
2.5	3.2	4.0	5.0	6.3	8.0	10.0	12.5

NOTE The rated hoisting speed of the wire rope hoist should be taken as 1.0 m/min to 2.5 m/min.

Appendix B Check of Motors

B.0.1 The motors of the lifting mechanism shall be subjected to overload check, and shall satisfy the following formula:

$$P_N \geq \frac{H}{m\lambda_m} \times \frac{(P_Q + q)v_q}{1000\eta} \tag{B.0.1}$$

where

P_N is the rated power of motor (kW);

H is the coefficient, which is determined based on the voltage loss (−15 % for AC motors, and zero for DC motor and variable frequency motor), maximum torque or locked-rotor torque tolerance (−10 % for wound-rotor asynchronous motors, −15 % for squirrel cage asynchronous motors, and zero for DC motors and variable frequency motors), and rated lifting load, and is taken as 2.5 for wound-rotor asynchronous motors and squirrel cage asynchronous motors, 2.2 for variable frequency asynchronous motors, and 1.4 for DC motors;

m is the number of motors;

λ_m is the multiple of the maximum torque of the motor relative to P_N, provided by the motor manufacturer. For the squirrel cage motor started directly at full voltage, the multiple of the locked-rotor torque, λ_m, is not less than 2.2;

P_Q is the rated lifting load (N);

q is the weight of slings and wire ropes (N);

v_q is the rated hoisting speed (m/s);

η is the overall efficiency of lifting mechanism.

B.0.2 The cyclic duration factor FC of the lifting mechanism is used for applications where the duty cycle is not less than 10 min and shall be calculated by the following formula:

$$FC = \frac{\text{Running time of the lifting mechanism in one duty cycle of the hoist}}{\text{Total time of one duty cycle of the hoist}} \times 100\% \tag{B.0.2}$$

B.0.3 CZ value calculation of wound-rotor asynchronous motor.

1 The rate of inertia increase C is calculated by the following formula:

$$C = \frac{GD_{d}^{2} + GD_{e}^{2}}{GD_{d}^{2}} \tag{B.0.3-1}$$

where

C is the rate of inertia increase;

GD_{d}^{2} is the flywheel moment of motor (kg · m²);

GD_{e}^{2} is the flywheel moment converted from the moving mass and rotating mass other than the motor to the motor shaft (kg · m²).

2 The converted number of complete starts per hour Z is calculated by the following formula:

$$Z = d_{0} + gd_{i} + rf \tag{B.0.3-2}$$

where

Z is the converted number of complete starts per hour;

d_{0} is the number of complete starts per hour;

d_{i} is the number of inching or incomplete starts per hour;

f is the number of electrical braking per hour;

g, r are the conversion factors, which may be taken as 0.25 and 0.25, respectively.

3 The CZ value, which is the product of the rate of inertia increase C and the converted number of complete starts per hour Z, is an important parameter that affects the heating of the motor in starting and braking states.

B.0.4 The average coefficient G for the steady-state load of the motor of the hoist lifting mechanism shall be calculated based on the actual load. If detailed information on the load is not available, it may be selected with reference to Table B.0.4.

Table B.0.4 Average coefficient *G* of steady-state load of motor

FC (%)	*CZ*	*G*
15	150	G_1
25	150	G_2
40	300	G_2
60	450	G_3

B.0.5 The steady-state average power of wound-rotor asynchronous motors shall be calculated by the following formula:

$$P_s = G\frac{(P_Q + q)v_q}{1000\eta} \tag{B.0.5}$$

where

P_s is the steady-state average power (kW);

G is the average coefficient for steady-state load; taken as 0.7 for G_1, 0.8 for G_2, and 0.9 for G_3; the grading of G shall be in accordance with Table B.0.4;

P_Q is the rated lifting load (N);

q is the weight of slings and wire ropes (N);

v_q is the rated hoisting speed (m/s);

η is the overall efficiency of lifting mechanism.

B.0.6 The allowable output power P of YZR series wound-rotor asynchronous motors at different FC values and different CZ values shall be selected from the motor product catalog. When P is greater than P_s, the motor heating is deemed qualified.

Appendix C Motor Power Correction Considering Working Environment

C.0.1 When the motor operates at an altitude exceeding 1000 m or the operating ambient temperature is inconsistent with the rated ambient temperature, the power correction of the motor is calculated by the following formula:

$$P'_{\mathrm{N}} = \frac{P_{\mathrm{N}}}{K} \tag{C.0.1}$$

where

P'_{N} is the corrected power used to select the motor based on the ambient temperature and altitude (kW);

P_{N} is the required motor power before correction (kW);

K is the power correction factor, determined by the power correction as a function of ambient temperature and altitude (Figure C.0.1).

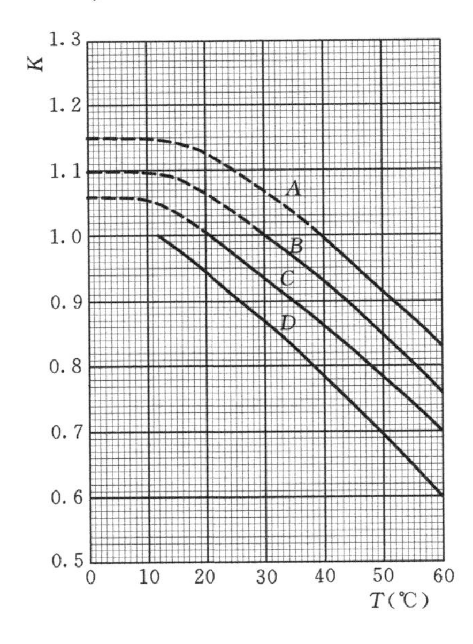

Key

K	power correction factor
T	ambient temperature
A	altitude below 1000 m
B	altitude below 2000 m
C	altitude below 3000 m
D	altitude below 4000 m

Figure C.0.1 Power correction as a function of ambient temperature and altitude

C.0.2 When the correction factor K is greater than 1, its value shall be jointly determined by the motor manufacturer and the hoist manufacturer.

C.0.3 The ambient temperature shall be indicated when the motor operates at an altitude exceeding 1000 m.

Appendix D Time and Acceleration/Deceleration for Starting/Braking the Lifting Mechanism

D.0.1 Starting time and average starting acceleration

1 The starting time of the lifting mechanism shall be calculated by the following formulae:

$$t_q = \frac{n\left[k(J_1 + J_2) + \frac{J_3}{\eta}\right]}{9.55(M_{dq} - M_N)} \tag{D.0.1-1}$$

$$J_3 = \frac{(P_Q + q)D^2}{4ga^2i^2} \tag{D.0.1-2}$$

$$M_{dq} = \lambda_{AS} M_n \tag{D.0.1-3}$$

where

t_q is the starting time of lifting mechanism (s);

n is the rated speed of motor (r/min);

k is the influence coefficient for the moment of inertia of other transmission parts converted to the motor shaft, taken as 1.05 to 1.20;

J_1 is the moment of inertia of motor rotor ($kg \cdot m^2$);

J_2 is the moment of inertia of brake wheel and coupling on the motor shaft ($kg \cdot m^2$);

J_3 is the moment of inertia of lifting moving objects converted to the motor shaft ($kg \cdot m^2$);

P_Q is the rated lifting load (N);

q is the weight of slings and wire ropes (N);

D is the pitch diameter (m);

g is the acceleration of gravity, taken as 9.81 m/s^2;

a is the ratio for pulley block system;

i is the total ratio from the motor shaft to the drum shaft;

η is the overall efficiency of the drive and pulley block of the lifting mechanism;

M_{dq} is the average starting torque of motor ($N \cdot m$);

λ_{AS} is the multiple of average starting torque of motor, which shall be in accordance with Table D.0.1;

M_n is the rated torque of motor (N · m);

M_N is the torque on the motor shaft during steady lifting at the rated load (N · m).

Table D.0.1 Multiple of average starting torque of motor

<table>
<tr><th colspan="2">Motor type</th><th>λ_{AS}</th></tr>
<tr><td colspan="2">Three-phase AC wound-rotor motor for hoist</td><td>1.5 - 1.8</td></tr>
<tr><td rowspan="2">Crane three-phase squirrel cage motor</td><td>Common type</td><td>Multiple of locked-rotor torque of motor</td></tr>
<tr><td>Frequency converter control type</td><td>1.5 - 1.8</td></tr>
</table>

2 The average starting acceleration is calculated by the following formula:

$$\alpha_q = \frac{v_q}{t_q} \tag{D.0.1-4}$$

where

α_q is the average starting acceleration of lifting mechanism (m/s^2);

v_q is the lifting speed (m/s);

t_q is the starting time of lifting mechanism (s), calculated by Formula (D.0.1-1).

D.0.2 Braking time and average braking deceleration

1 When the mechanical brake is used, the braking time for lowering at full load is calculated by the following formulae:

$$t_Z = \frac{n'\left[k\left(J_1 + J_2\right) + J_3\eta\right]}{9.55\left(M_Z - M_j'\right)} \tag{D.0.2-1}$$

$$M_j' = \frac{\left(P_Q + q\right)D}{2ai}\eta' \tag{D.0.2-2}$$

where

t_Z is the braking time of lifting mechanism (s);

n' is the motor speed when the effective brake torque is put into use during rated-load lowering (r/min), taken as $1.1n$, where n is the same as that in Formula (D.0.1-1);

k is the influence coefficient for the moment of inertia of other transmission parts converted to the motor shaft, taken as 1.05 to 1.20;

J_1 is the moment of inertia of motor rotor (kg · m^2);

J_2 is the moment of inertia of brake wheel and coupling on the motor shaft (kg · m^2);

J_3 is the moment of inertia of lifting moving objects converted to the motor shaft (kg · m^2), calculated by Formula (D.0.1-2);

η is the overall efficiency of the drive and pulley block of the lifting mechanism;

M_Z is the calculation braking torque of mechanical brake (N · m);

M_j' is the torque on the motor brake shaft during steady lowering at the rated load (N · m);

P_Q is the rated lifting load (N);

q is the weight of slings and wire ropes (N);

D is the pitch diameter (m);

a is the ratio for pulley block system;

i is the total ratio between the brake shaft and the drum shaft;

η' is the overall efficiency of the lifting mechanism during lowering the load.

2 The average braking deceleration is calculated by the following formula:

$$\alpha_Z = \frac{v_q'}{t_Z} \qquad \text{(D.0.2-3)}$$

where

α_Z is the average braking deceleration (m/s^2);

v_q' is the lowering speed when the brake starts acting effectively during full-load lowering (m/s), which may be taken as $1.1v_q$, where v_q is the same as that in Formula (D.0.1-4);

t_Z is the braking time of lifting mechanism (s), calculated by Formula (D.0.2-1).

Appendix E Calculation for Parts and Components

E.1 Shaft

E.1.1 The shaft shall be subjected to stress calculation based on the stressing conditions, which shall meet the following requirements:

1 The stress of shaft subjected to bending moment shall be calculated by Formula (E.1.1-1) and shall satisfy Formula (E.1.1-2):

$$\sigma_w = \frac{32DM_w}{\pi\left(D^4 - d^4\right)} \tag{E.1.1-1}$$

$$\sigma_w \leq [\sigma_w] \tag{E.1.1-2}$$

2 The stress of shaft subjected to torsion shall be calculated by Formula (E.1.1-3) and shall satisfy Formula (E.1.1-4):

$$\tau_n = \frac{16DM_n}{\pi\left(D^4 - d^4\right)} \tag{E.1.1-3}$$

$$\tau_n \leq [\tau_n] \tag{E.1.1-4}$$

3 The stress of shaft subjected to the combined action of bending and torsion shall be calculated by Formula (E.1.1-5) and shall satisfy Formula (E.1.1-6):

$$\sigma_h = \sqrt{\sigma_w^2 + 4\tau_n^2} \tag{E.1.1-5}$$

$$\sigma_h \leq [\sigma_w] \tag{E.1.1-6}$$

where

σ_w is the calculated bending stress of shaft (N/mm^2);

D is the outer diameter of the shaft at the calculation section (mm);

M_w is the bending moment of the shaft at the calculation section (N · mm);

d is the inner diameter of the shaft at the calculation section (mm), taken as 0 for a solid shaft;

$[\sigma_w]$ is the allowable bending stress of shaft (N/mm^2);

τ_n is the calculated torsional stress of shaft (N/mm^2);

M_n is the torsional moment of the shaft at the calculation section (N · mm);

$[\tau_n]$ is the allowable torsional stress of shaft (N/mm^2);

σ_h is the calculated resultant bending-torsional stress of shaft (N/mm^2).

E.1.2 The allowable bending stress and allowable torsional stress of the shaft shall meet the following requirements:

1 The allowable bending stress of the shaft shall be calculated by formula (E.1.2-1):

$$[\sigma_w]=\frac{R_{eL}}{n} \tag{E.1.2-1}$$

where

n is the safety factor, taken as at least 2.5 for lifting shafts and hinged shafts, 3.0 for pulley shafts, and 4.0 for drum shafts.

2 The allowable torsional stress of the shaft shall be calculated by formula (E.1.2-2):

$$[\tau_n] = 0.6[\sigma_w] \tag{E.1.2-2}$$

E.2 Lifting plate

The strength calculation of the lifting plate structure (Figure E.2.1-1) shall meet the following requirements:

1 The bearing stress of the lifting plate hole wall shall be calculated by Formula (E.2.1-1) and shall satisfy Formula (E.2.1-2):

$$\sigma_m=\frac{P}{d\delta} \tag{E.2.1-1}$$

$$\sigma_m\leq\frac{R_{eH}}{n} \tag{E.2.1-2}$$

where

σ_m is the compressive stress of lifting plate hole wall (N/mm^2);

P is the load borne by one lifting plate (N);

d is the diameter of lifting plate shaft hole (mm);

δ is the thickness of lifting plate (mm);

R_{eH} is the upper yield limit of steel used for lifting plate (N/mm^2);

n is the safety factor, taken as 5.0 for the shaft hole with relative rotation to the shaft and 3.0 for the shaft hole without relative rotation to the shaft.

2 The tensile stress of the horizontal section of the lifting plate shaft hole shall be calculated by Formula (E.2.1-3) and shall satisfy Formula (E.2.1-4):

$$\sigma_1 = \frac{P}{(B-d)\delta} a_j \tag{E.2.1-3}$$

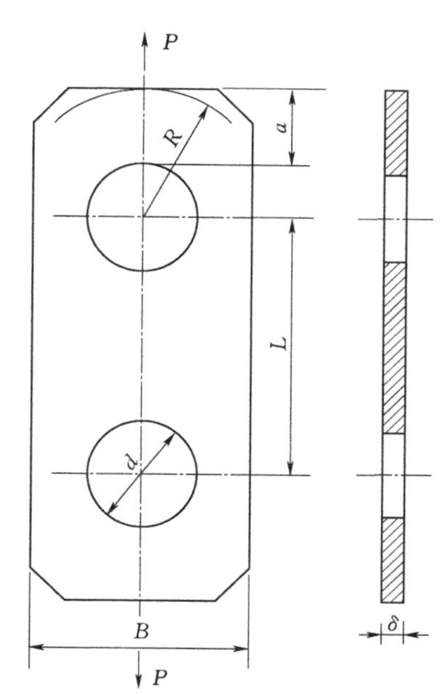

Key

B width of lifting plate structure

d diameter of lifting plate shaft hole

δ thickness of lifting plate structure

a minimum section height from the outer edge of the shaft hole to the outer edge of the lifting plate

L center distance between shaft holes

Figure E.2.1-1 Schematic diagram of lifting plate structure

$$\sigma_1 \leq \frac{R_{eH}}{1.7} \tag{E.2.1-4}$$

where

σ_1 is the tensile stress on horizontal section of lifting plate shaft hole (N/mm^2);

P is the load borne by one lifting plate (N);

B is the width of lifting plate (mm);

d is the diameter of lifting plate shaft hole (mm);

δ is the thickness of lifting plate (mm);

a_j is the stress concentration factor (Figure E.2.1-2);

R_{eH} is the upper yield limit of steel used for lifting plate (N/mm^2).

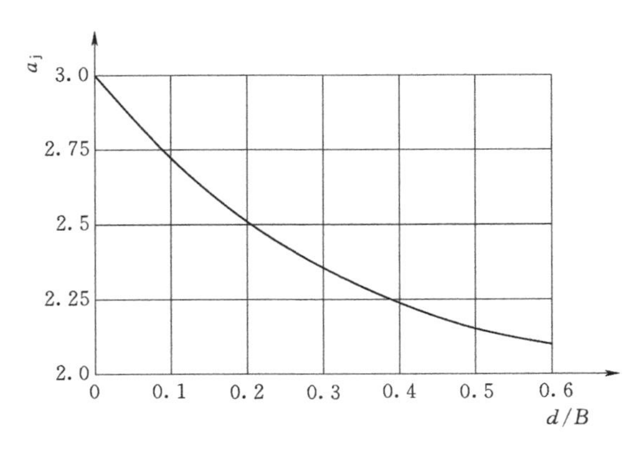

Figure E.2.1-2 Stress concentration factor

3 The tensile stress on the vertical section of the lifting plate shaft hole shall be calculated by Formula (E.2.1-5) and shall satisfy Formula (E.2.1-6):

$$\sigma_e = \frac{P\left(R^2 + 0.25d^2\right)}{d\delta\left(R^2 - 0.25d^2\right)} \tag{E.2.1-5}$$

$$\sigma_e \leq \frac{R_{eH}}{3} \tag{E.2.1-6}$$

where

σ_e is the tensile stress on the vertical section of lifting plate shaft hole (N/mm^2);

P is the load borne by one lifting plate (N);

R is the smaller of $B/2$ or $(d/2+a)$;

d is the diameter of lifting plate shaft hole (mm);

δ is the thickness of lifting plate (mm);

R_{eH} is the upper yield limit of steel used for lifting plate (N/mm^2).

E.3 Drum

E.3.1 The calculation of the drum wall shall meet the following requirements:

1 When the drum length L is not greater than 3 times the drum diameter D, the compressive stress in the drum wall shall be calculated by Formula (E.3.1-1) and shall satisfy Formula (E.3.1-2). When the drum is made of steel plate or cast steel, Formula (E.3.1-3) shall be satisfied. When the drum is made of cast iron, Formula (E.3.1-4) shall be satisfied.

$$\sigma_p = A\frac{S}{\delta t} \tag{E.3.1-1}$$

$$\sigma_p \leq [\sigma_p] \tag{E.3.1-2}$$

$$[\sigma_p] = \frac{R_{eL}}{1.5} \tag{E.3.1-3}$$

$$[\sigma_p] = \frac{R_c}{4.25} \tag{E.3.1-4}$$

where

L is the length of drum (mm);

D is the tread diameter of drum (mm);

σ_p is the compressive stress of drum wall (N/mm^2);

A is the multi-layer winding coefficient related to the number of winding layers of the wire rope, selected as per Table E.3.1;

S is the maximum working tension of wire rope (N);

δ is the wall thickness of drum (mm), taken as at least 12 mm for cast iron and at least 15 mm for cast steel;

t is the pitch of rope groove on the drum (mm);

$[\sigma_p]$ is the allowable compressive stress (N/mm^2);

R_{eL} is the yield strength (N/mm^2);

R_c is the compressive strength (N/mm^2).

Table E.3.1 Multi-layer winding coefficient

Number of winding layers	1	2	3	≥4
***A* value**	1.0	1.4	1.8	2

2 When the drum length L is greater than 3 times the drum diameter D, the conversion stress generated by the bending moment and torsional moment shall be calculated by Formula (E.3.1-5) and Formula (E.3.1-6) and shall satisfy Formula (E.3.1-7). When the drum is made of plate steel or cast steel, Formula (E.3.1-8) shall be satisfied. When the drum is made of cast iron, Formula (E.3.1-9) shall be satisfied.

$$\sigma_F = M_F / W \tag{E.3.1-5}$$

$$M_F = \sqrt{M_w^2 + M_n^2} \tag{E.3.1-6}$$

$$\sigma_F \leq [\sigma] \tag{E.3.1-7}$$

$$[\sigma]=\frac{R_{eL}}{2.5} \quad \text{(E.3.1-8)}$$

$$[\sigma]=\frac{R_m}{6} \quad \text{(E.3.1-9)}$$

where

σ_F is the conversion stress (N/mm^2);

M_F is the conversion moment (N/mm^2);

W is the resistance moment of drum section (mm^3);

M_w is the bending moment of the drum (N · mm);

M_n is the torque of the drum (N · mm);

$[\sigma]$ is the allowable stress (N/mm^2);

R_{eL} is the yield strength (N/mm^2);

R_m is the tensile strength (N/mm^2).

3 When the drum diameter D is greater than or equal to 1,200 mm, or the drum length L is greater than 2 times the drum diameter D, the stability of the drum shall be calculated by Formula (E.3.1-10) and Formula (E.3.1-11) and shall satisfy Formula (E.3.1-12). When the drum is made of steel plate or cast steel, Formula (E.3.1-13) shall be satisfied. When the drum is made of cast iron, Formula (E.3.1-14) shall be satisfied.

$$K=p_w/p \quad \text{(E.3.1-10)}$$

$$p=\frac{2S}{tD} \quad \text{(E.3.1-11)}$$

$$K\geq 1.5 \quad \text{(E.3.1-12)}$$

$$p_w=\frac{52500\delta^3}{R^3} \quad \text{(E.3.1-13)}$$

$$p_w=\frac{25000\delta^3}{R^3} \quad \text{(E.3.1-14)}$$

where

K is the stability coefficient;

p_w is the critical stress of stability (N/mm^2);

δ is the wall thickness of drum (mm), taken as at least 12 mm for cast iron and at least 15 mm for cast steel;

R is the radius of rope groove bottom of drum (mm), taken as $D/2$;

p is the unit compressive stress of drum wall (N/mm^2);

S is the maximum working tension of the wire rope corresponding to the calculation layer (N);

t is the pitch of rope groove on the drum (mm).

E.3.2 The calculation of the drum shaft shall meet the following requirements:

1 For the long-rotary shaft used in double-reeved drum shaft with open gear, the horizontal bending moment, vertical bending moment and corresponding bending stress of each section shall be calculated according to the Schematic Diagram of Force Calculation of Double-Reeved Drum Shaft with Open Gear (Figure E.3.2-1).

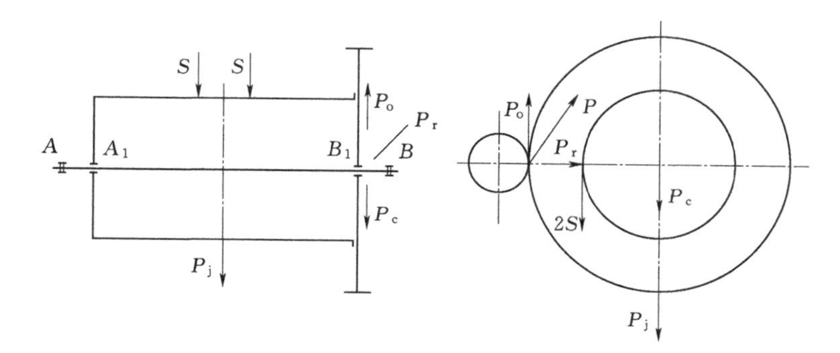

(a) Schematic diagram of structure layout and forces on drum shaft

(b) Schematic diagram of forces in side-view

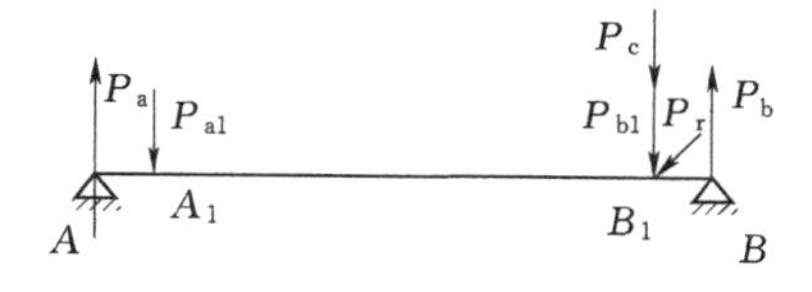

(c) Calculation diagram

Key

S maximum static working tension of wire rope

P_j gravity of drum, wire rope and drum shaft

P_c gravity of open gear

P_o circumferential force of large gear

P_r radial force of large gear

P calculated resultant force of P_o and P_r

P_{a1}, P_{b1} reaction forces under the action of S and P_j

P_a, P_b Reaction forces of seat bearing

Figure E.3.2-1 Schematic diagram of force calculation of double-reeved drum shaft with open gear

2 For the short-welded shaft used in double-reeved drums driven with closed gear, the bending moment and bending stress of each section

of the shaft shall be calculated according to the Schematic Diagram of Force Calculation of Welded Short Drum Shaft (Figure E.3.2-2).

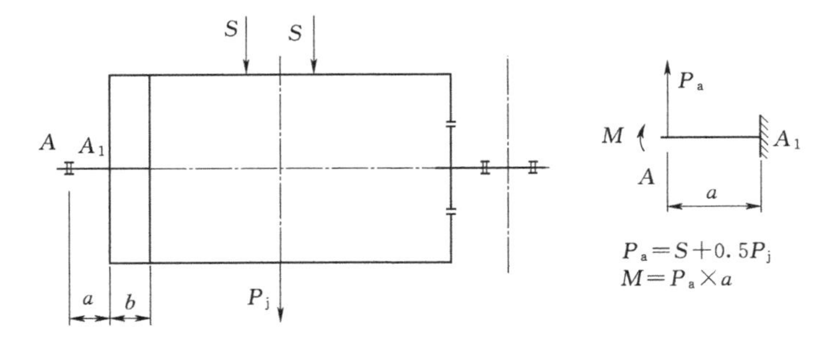

(a) Schematic diagram of structure layout and force of welded short drum shaft

(b) Calculation diagram

Key

S maximum static working tension of wire rope

P_j gravity of drum, wire rope and drum shaft

P_a reaction force of seat bearing

Figure E.3.2-2 Schematic diagram of force calculation of welded short drum shaft

3 For the short shaft connected to drum with sleeve, and used in double-reeved drums driven by closed gear, the bending moment and bending stress of each section of the shaft shall be calculated according to Schematic Diagram of Force Calculation of Short Drum Shaft Connected by a Sleeve (Figure E.3.2-3).

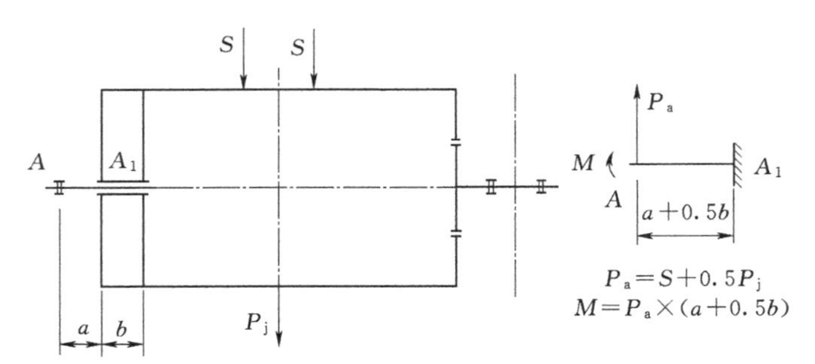

(a) Schematic diagram of structure layout and force of sleeve connected short shaft of drum

(b) Calculation diagram

Key

S maximum working static tension of wire rope

P_j gravity of drum, wire rope and drum shaft

P_a reaction force of seat bearing

Figure E.3.2-3 Schematic diagram of force calculation of short drum shaft connected by a sleeve

E.3.3 The calculation of connection of large gear and drum shall meet the following requirements:

1 When the large gear is connected to the drum by sleeves (Figure E.3.3), the torque is transmitted by the sleeve, and the connecting bolts are not subjected to shear and only play the role of connection. The shear stress of the sleeve shall be calculated by Formula (E.3.3-1) and shall satisfy Formula (E.3.3-2):

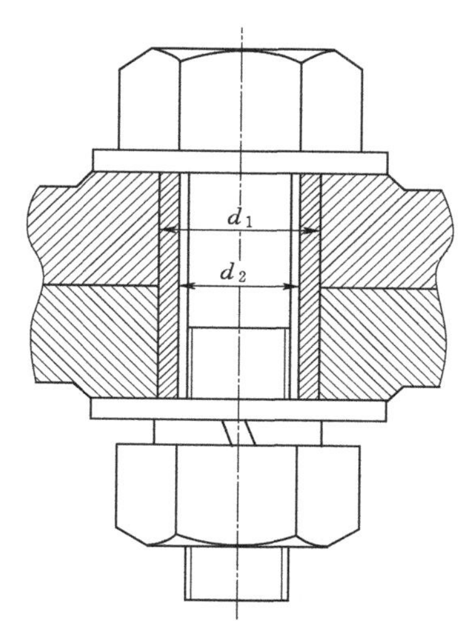

Figure E.3.3 Connection of drum and large gear by sleeves

$$\tau = \frac{8M}{n\pi\left(d_1^2 - d_2^2\right)D} \tag{E.3.3-1}$$

$$\tau \leq [\tau] \tag{E.3.3-2}$$

where

τ is the shear stress of sleeve (N/mm^2);

M is the torque transmitted by sleeve (N · mm);

n is the number of sleeves;

d_1 is the outer diameter of sleeve (mm);

d_2 is the inner diameter of sleeve (mm);

D is the diameter of central circle formed by sleeves (mm);

$[\tau]$ is the allowable shear stress of sleeve (N/mm^2), taken as 85 N/mm^2 for the sleeve made of Grade 45 steel subjected to heat treatment.

2 When the large gear is connected to the drum by reamed hole bolts, the torque is directly transmitted by the bolts which also play the role of connection. The shear stress of the reamed hole bolt shall be calculated by Formula (E.3.3-3) and shall satisfy Formula (E.3.3-4):

$$\tau = \frac{8M}{n\pi d_1^2 D} \tag{E.3.3-3}$$

$$\tau \leq [\tau] \tag{E.3.3-4}$$

where

τ is the shear stress of reamed hole bolt (N/mm^2);

M is the torque transmitted by reamed hole bolt (N · mm);

n is the number of reamed hole bolts;

d_1 is the diameter of polished rod part of reamed hole bolt (mm);

D is the diameter of central circle formed by reamed hole bolts (mm);

$[\tau]$ is the allowable shear stress of reamed hole bolt (N/mm^2).

3 Based on the length L of the sleeve or reamed hole bolt where the torque is transmitted, the extrusion stress shall be calculated by Formula (E.3.3-5) and shall satisfy Formula (E.3.3-6):

$$\sigma_{cm} = \frac{2M}{nd_1 LD} \tag{E.3.3-5}$$

$$\sigma_{cm} \leq [\sigma_{cm}] \tag{E.3.3-6}$$

where

σ_{cm} is the extrusion stress between the sleeve or reamed hole bolt and the drum or large gear (N/mm^2);

M is the torque transmitted by the sleeve or reamed hole bolt (N · mm);

n is the number of sleeves or reamed hole bolts;

d_1 is the outer diameter of sleeve or the diameter of polished rod part of reamed hole bolt (mm);

L is the length of the sleeve or reamed hole bolt where the torque is transmitted (mm);

D is the diameter of central circle formed by sleeves or reamed hole bolts (mm);

$[\sigma_{cm}]$ is the allowable extrusion stress (N/mm^2).

E.3.4 The calculation of clamping plate bolts shall meet the following requirements:

1 When the wire rope is fixed by the clamping plates and bolts (Figure E.3.4), the tensile stress of the clamping plate bolt includes the tensile stress caused by the pressing force and the tensile stress caused by the bending of the bolt due to the friction between the wire rope and the clamping plate.

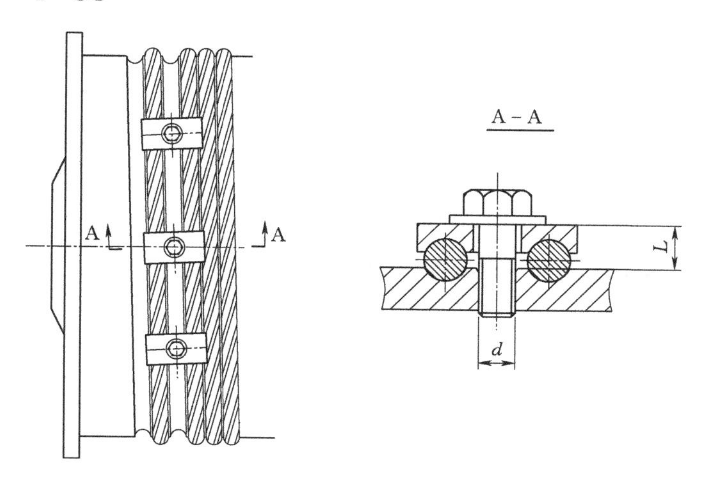

Figure E.3.4 Fixing of wire rope by clamping plates and bolts

2 The tensile force at the fixation place of the wire rope is calculated by the following formula:

$$S=\frac{\varphi_z S_{max}}{e^{\mu\alpha}} \quad \text{(E.3.4-1)}$$

where

S is the tensile force at the fixation place of the wire rope (N);

φ_z is the dynamic load factor of lifting load, taken as 1.0 to 1.1 for operating gate in still water and 1.1 to 1.2 for operating gate in flowing water;

S_{max} is the maximum static tension of wire rope (N);

e is the base of the natural logarithm, taken as 2.718282;

μ is the coefficient of friction between the wire rope and the drum or clamping plate, taken as 0.12 to 0.16 depending on whether there is grease on the faying surface of the steel plate-rolled or cast steel drum;

α is the wrap angle of safety wrap on the drum (rad), taken as 4π

for two safety wraps and 6π for three safety wraps.

3 The pressing force of a single circular clamping plate bolt on the wire rope is calculated by the following formula:

$$N = \frac{n_0 S}{2\mu Z} \tag{E.3.4-2}$$

where

N is the pressing force of clamping plate bolt on wire rope (N);

n_0 is the safety factor, generally taken as 1.5 or larger;

S is the same as that in Formula (E.3.4-1);

μ is the same as that in Formula (E.3.4-1);

Z is the number of bolts, taken as 2 or larger.

4 The tensile stress of the circular clamping plate bolt shall be calculated by Formula (E.3.4-3) and shall satisfy Formula (E.3.4-4):

$$\sigma = \frac{4N}{\pi d^2} + \frac{10\mu NL}{d^3} \tag{E.3.4-3}$$

$$\sigma \leq [\sigma] \tag{E.3.4-4}$$

where

σ is the tensile stress of clamping plate bolt (N/mm^2);

N is the pressing force of clamping plate on wire rope, same as that in Formula (E.3.4-2);

d is the inner diameter of thread for fixing bolt (mm);

μ is the same as that in Formula (E.3.4-1);

L is the force arm of wire rope tension on bolt root (mm);

$[\sigma]$ is the allowable tensile stress of bolt (N/mm^2), taken as per Table 8.3.3-1.

5 The strength of the clamping plate shall be checked according to the pressing force of the clamping plate bolt.

Appendix F Allowable Values of Common Sliding Bearing Materials

F.0.1 The maximum allowable values of copper alloy bearing materials should be taken as per Table F.0.1.

Table F.0.1 Maximum allowable values of copper alloy bearing materials

Material of faying surface		[*p*] (N/mm^2)	[*v*] (m/s)	[*pv*] (N · m)/(mm^2 · s)
Tin bronze	ZCuSn10P1	15	10	15
	ZCuSn5Pb5Zn5	8	6	6
Cast aluminum bronze	ZCuAl10Fe3	30	8	12
	ZCuAl10Fe3Mn2	20	5	15
Cast lead bronze	ZCuPb30	15	8	60

F.0.2 Manufacturers of self-lubricating bearing materials such as steel-based copper-plastic composite materials, copper alloy inlaid self-lubricating materials, and engineering plastic alloy materials shall entrust the testing agencies with professional qualifications to determine [*p*], [*v*] and [*pv*] by the test methods specified in relevant technical codes of China.

Appendix G Correction of Wire Ampacity

G.0.1 The temperature correction factor for wire ampacity is calculated by the following formula, and the common value may be selected as per Table G.0.1.

$$K_t = \sqrt{\frac{T_1 - T_0}{T_1 - T_2}} \tag{G.0.1}$$

where

K_t is the ambient temperature correction factor;

T_1 is the maximum operating temperature of wire core (°C);

T_0 is the operating ambient temperature (°C);

T_2 is the rated operating ambient temperature (°C), taken as 25 °C or 45 °C.

G.0.2 The wire ampacity is calculated by the following formulae:

$$I_z = K_a K_t K_j I_g \tag{G.0.2-1}$$

$$K_j = \sqrt{\frac{1 - e^{\frac{-600}{T}}}{1 - e^{\frac{-600FC}{T}}}} \tag{G.0.2-2}$$

where

I_z is the wire ampacity (A);

K_a is the correction factor for parallel laying of multiple cables or in-conduit wires, which may be taken as 0.9 for in-conduit wires and 0.8 for cables;

K_t is the ambient temperature correction factor;

K_j is the correction factor for cyclic duration factor for repeated short-time duty, the common value of which may be selected as per Table G.0.2-1 when the duty cycle is 10 min;

I_g is the reference value of ampacity of wire or cable, provided by the cable manufacturer, the common value of which may be selected as per Table G.0.2-2;

T is the wire thermal time constant (s), provided by the cable manufacturer, the common value of which may be selected as per Table G.0.2-2;

FC is the cyclic duration factor.

Table G.0.1 Temperature correction factor K_t for wire ampacity

Rated operating ambient temperature (° C)	Maximum operating temperature of wire core (° C)	Operating ambient temperature (° C)										
		+25	+30	+35	+40	+45	+50	+55	+60	+65	+70	+75
+25	+60	1.000	0.926	0.845	0.756	0.655	0.535	–	–	–	–	–
	+65	1.000	0.935	0.865	0.791	0.707	0.612	0.500	–	–	–	–
	+70	1.000	0.943	0.882	0.816	0.745	0.667	0.577	0.471	–	–	–
+45	+65	–	1.323	1.225	1.118	1.000	0.866	0.707	–	–	–	–
	+70	–	1.265	1.183	1.095	1.000	0.894	0.775	0.632	–	–	–
	+80	–	1.195	1.134	1.069	1.000	0.926	0.845	0.756	0.655	0.535	–
	+85	–	1.173	1.118	1.061	1.000	0.835	0.866	0.791	0.707	0.612	0.500

Table G.0.2-1 Correction factor K_j for cyclic duration factor of repeated short-time duty

Wire type	Cyclic duration factor	Cross-sectional area of wire core (mm^2)												
		1.5	2.5	4	6	10	16	25	35	50	70	95	120	150
BX or BXR copper core rubber-sheathed wire	25 %	1.313	1.417	1.477	1.550	1.614	1.678	1.754	1.790	1.834	1.849	1.876	1.880	1.898
	40 %	1.149	1.212	1.249	1.296	1.336	1.377	1.425	1.448	1.476	1.486	1.503	1.505	1.517
YC, YCW, CF or CFR single-core cable	25 %	1.250	1.304	1.324	1.398	1.461	1.520	1.604	1.645	1.701	1.742	1.784	1.807	1.830
	40 %	1.111	1.143	1.155	1.200	1.240	1.277	1.330	1.356	1.391	1.417	1.444	1.459	1.473
YC, YCW, CF or CFR three-core cable	25 %	1.490	1.531	1.590	1.640	1.696	1.750	1.808	1.823	1.838	1.865	1.877	1.902	1.912
	40 %	1.258	1.284	1.321	1.353	1.388	1.422	1.460	1.456	1.479	1.495	1.510	1.519	1.526

Table G.0.2-2 Wire ampacity reference value and wire thermal time constant

Cross-sectional area of wire core (mm^2)	Copper-core wire						Heavy rubber-sheathed cable				Marine cable			
	BX or BXR copper core rubber-sheathed wire		BV or BVR copper-core plastic wire		Thermal time constant (s)		YC or YCW single-core cable		YC or YCW three-core cable		CF or CFR single-core cable		CF or CFR three-core cable	
	Ampacity at 25 °C (A)						Ampacity at 25 °C (A)	Thermal time constant (s)	Ampacity at 25 °C (A)	Thermal time constant (s)	Ampacity at 25 °C (A)	Thermal time constant (s)	Ampacity at 25 °C (A)	Thermal time constant (s)
	Exposed	In-conduit	Exposed	In-conduit	Exposed	In-conduit								
1.5	27	18	24	17	86	184	-	-	-	-	20	152	14	307
2.5	35	25	32	24	116	248	37	179	26	347	26	179	19	347
4	45	33	42	31	138	295	47	190	34	419	35	190	25	419
6	58	43	55	41	172	368	52	235	43	497	44	235	32	497
10	185	60	75	57	212	453	75	282	63	613	61	282	44	613
16	110	77	105	73	267	571	112	336	84	774	81	336	58	774
25	145	100	138	95	370	791	148	438	115	1050	105	438	77	1050
35	180	122	170	115	442	945	183	506	142	1020	135	506	94	1020
50	230	154	215	146	573	1230	226	626	176	1270	165	626	120	1270
70	285	193	265	183	641	1370	289	746	224	1540	205	746	145	1540
95	345	235	325	225	797	1700	353	917	273	1870	250	917	180	1870
120	400	270	375	260	820	1750	415	1040	316	2180	290	1040	205	2180
150	470	310	430	300	980	2090	-	-	-	-	335	1200	240	2450

NOTE The ampacity of the in-conduit wire in the table is based on three single-core wires laid in the steel conduit in the open air. For simplicity, the cross-sectional area of wire core for all wires of the hoist, regardless of the laying method and location, is selected based on the condition that three single-core wires are generally laid in a steel conduit in the open air. When the number of wires in a conduit exceeds three, the ampacity may be appropriately reduced for the selected cross-sectional area.

Explanation of Wording in This Code

1 Words used for different degrees of strictness are explained as follows in order to mark the differences in executing the requirements in this code:

1) Words denoting a very strict or mandatory requirement:

"Must" is used for affirmation; "must not" for negation.

2) Words denoting a strict requirement under normal conditions:

"Shall" is used for affirmation; "shall not" for negation.

3) Words denoting a permission of a slight choice or an indication of the most suitable choice when conditions permit:

"Should" is used for affirmation; "should not" for negation.

4) "May" is used to express the option available, sometimes with the conditional permit.

2 "Shall meet the requirements of..." or "shall comply with..." is used in this code to indicate that it is necessary to comply with the requirements stipulated in other relative standards and codes.

List of Quoted Standards

GB/T 699,	*Quality Carbon Structural Steels*
GB/T 700,	*Carbon Structural Steels*
GB/T 985.1,	*Recommended Joint Preparation for Gas Welding, Manual Metal Arc Welding, Gas-Shield Arc Welding and Beam Welding*
GB/T 985.2,	*Recommended Joint Preparation for Submerged Arc Welding*
GB/T 1176,	*Cast Copper and Copper Alloys*
GB/T 1231,	*Specifications of High Strength Bolts with Large Hexagon Head, Large Hexagon Nuts, Plain Washers for Steel Structures*
GB/T 1348,	*Spheroidal Graphite Iron Castings*
GB/T 1591,	*High Strength Low Alloy Structural Steels*
GB 2893,	*Safety Colours*
GB 2894,	*Safety Signs and Guideline for the Use*
GB/T 3077,	*Alloy Structure Steels*
GB/T 3098.1,	*Mechanical Properties of Fasteners—Bolts, Screws and Studs*
GB/T 3098.2,	*Mechanical Properties of Fasteners —Nuts*
GB/T 3098.3,	*Mechanical Properties of Fasteners—Set Screws*
GB/T 3098.6,	*Mechanical Properties of Fasteners — Stainless Steel Bolts, Screws and Studs*
GB/T 3098.15,	*Mechanical Properties of Fasteners—Stainless Steel Nuts*
GB/T 3480,	*Calculation Methods of Load Capacity for Involute Cylindrical Gears*
GB/T 3632,	*Sets of Torshear Type High Strength Bolt Hexagon Nut and Plain Washer for Steel Structures*
GB/T 3811,	*Design Rules for Cranes*
GB/T 4208,	*Degrees of Protection Provided by Enclosure (IP Code)*
GB/T 4942.1,	*Degrees of Protection Provided by the Integral Design of*

	Rotating Electrical Machines (IP Code)—Classification
GB/T 5117,	*Covered Electrodes for Manual Metal Arc Welding of Non-alloy and Fine Grain Steels*
GB/T 5118,	*Covered Electrodes for Manual Metal Arc Welding of Creep-Resisting Steels*
GB/T 5226.32,	*Electrical Safety of Machinery—Electrical Equipment of Machines—Part 32: Requirements for Hoisting Machines*
GB/T 5293,	*Solid Wire Electrodes, Tubular Cored Electrodes and Electrode/Flux Combinations for Submerged Arc Welding of Non Alloy and Fine Grain Steels*
GB/T 5975,	*Clamping Plates for Fixing Steel Wire Ropes*
GB/T 6067.1,	*Safety Rules for Lifting Appliances—Part 1: General*
GB/T 8110,	*Welding Electrodes and Rods for Gas Shielding Arc Welding of Carbon and Low Alloy Steel*
GB/T 8918,	*Steel Wire Ropes for Important Purposes*
GB/T 9439,	*Grey Iron Castings*
GB/T 10045,	*Tubular Cored Electrodes for Non-alloy and Fine Grain Steels*
GB/T 11352,	*Carbon Steel Castings for General Engineering Purpose*
GB/T 12470,	*Solid Wire Electrodes, Tubular Cored Electrodes and Electrode/Flux Combinations for Submerged Arc Welding of Creep-Resisting Steels*
GB/T 14048.4,	*Low-voltage Switchgear and Controlgear—Part 4-1: Contactors and Motor-Starters—Electromechanical Contactors and Motor-Starters (Including Motor Protector)*
GB/T 14957,	*Steel Wires for Melt Welding*
GB/T 15052,	*Cranes—Safety Signs and Hazard Pictorials—General Principles*
GB/T 17493,	*Tubular Cored Electrodes for Creep-Resisting Steels*
GB/T 17853,	*Tubular Cored Electrodes for Stainless Steels*
GB/T 17909.1,	*Cranes—Crane Driving Manual —Part 1: General*
GB/T 18453,	*Cranes—Maintenance Manual—Part 1: General*

GB/T 27546,	*Sheaves for Cranes*
GB/T 33083,	*Heavy Carbon Structural Steel Forgings—Technical Specification*
GB/T 33084,	*Heavy Alloy Structural Steel Forgings—Technical Specification*
GB 50009,	*Load Code for the Design of Building Structures*
GB 50017,	*Standard for Design of Steel Structures*
GB 50872,	*Code for Fire Protection Design of Hydropower Projects*
DL/T 5358,	*Technical Code for Anticorrosion of Metal Structures in Hydroelectric and Hydraulic Engineering*
JB/T 6402,	*Heavy Low Alloy Steel Castings—Technical Specification*
JB/T 6406,	*Electro-Hydraulic Drum Brakes*
JB/T 6398,	*Heavy Stainless Acid Resistant Steel and Heat Resistant Steel Forgings—Technical Specification*
JB/T 7020,	*Electro-Hydraulic Disc Brakes*
JB/T 7685,	*Electro-Magnetic Drum Brakes*
JB/T 9006,	*Drums for Cranes*
JGJ 82,	*Technical Specification for High Strength Bolt Connections of Steel Structures*
NB/T 35051,	*Code for Manufacture Erection and Acceptance of Gate Hoists in Hydropower Projects*
YB/T 5092,	*Stainless Steel Wires for Welding*
YB/T 5359,	*Compacted Strand Rope*